California Civil Surveying Practice Exams

Fourth Edition

Peter R. Boniface, PhD, PLS, and
Allan Y. Ng, MSCE, PLS

1 in = 2.54 cm

1 mile = 5280 feet.

1 acre = 43 560 ft²

1 yard = 3 ft

PPI®

PPI2PASS.COM

A **KAPLAN** COMPANY

Report Errors for This Book

PPI is grateful to every reader who notifies us of a possible error. Your feedback allows us to improve the quality and accuracy of our products. Report errata at **ppi2pass.com**.

CALIFORNIA CIVIL SURVEYING PRACTICE EXAMS
Fourth Edition

Current release of this edition: 3

Release History

date	edition number	revision number	update
Jan 2019	4	1	New edition.
Jun 2019	4	2	Minor cover updates.
Feb 2022	4	3	Minor corrections.

PPI
ppi2pass.com

ISBN: 978-1-59126-641-9

Table of Contents

Preface and Acknowledgments

California Civil Surveying Practice Exams is designed to simulate the exam experience and give you the authentic preparation you need for the California Civil Engineering Surveying Exam. The problems in this book are based on the Engineering Surveying Test Plan as published on the website of the California Board for Professional Engineers, Land Surveyors, and Geologists. The exam syllabus is extremely broad and touches on topics that most civil engineers have not studied during their undergraduate education. This is why two sample exams are provided in this book—so you can double your problem-solving practice.

The 55 problems in each sample exam mirror the actual exam problems in subject matter, length, and degree of difficulty. Although past examinations are not generally made publicly available, the published test plan does list all topics that appear on the exam and gives the following percentages for the coverage of groups of topics: topographic surveys, 35%; construction surveys, 35%; accuracy and error analysis, 10%; and preparation of reports and maps, 20%.

California Civil Surveying Practice Exams contains sample exams that are in alignment with the California Civil Engineering Surveying Exam, which has changed from a paper-based exam to a computer-based one. The sample exams include both simple and complex problems, which will prepare you more fully for the experience of taking the actual exam. Additionally, the problems reflect the scope of topics that the exam tests for, each represented in proportion to the number of problems that can be expected on the exam. We feel that this edition is an excellent resource to help you prepare for the exam.

We would like to thank Michael Lee, PE, PLS, and George M. Cole, Jr., PE, PLS, for performing the technical reviews of the first and second editions of this book, respectively; Elena Badilla, PE, and Alfredo Cely, PE, validity reviewers; Allan Y. Ng, MSCE, PLS, for validity reviewing and co-authoring this book; Joshua T. Frohman, PE, problem writer and contributor; and Ralph Arcena, calculations checker. We would also like to thank the PPI editorial and production staffs including Meghan Finley, content specialist; Steve Shea, product manager; Sam Webster, product data manager; Ellen Nordman, publishing systems specialist; Beth Christmas, editorial project manager; Scott Rutherford, copy editor; Tom Bergstrom, technical drawings and cover design; Cathy Schrott, production services manager; and Grace Wong, director of publishing services, for their hard work and patience in helping to put this edition together. Finally, we would like to thank James R. Monroe, Jr., PE, for writing *Practice Exams for the California Civil Surveying Exam*, from which this book is derived.

Peter R. Boniface, PhD, PLS

Introduction

This book is designed to prepare you for the California Civil Engineering Surveying Exam. California requires candidates for licensure as civil engineers to pass the exam, among other registration requirements.

You should begin your exam preparation by reviewing a wide range of textbooks in this field. You will need a solid understanding of the fundamentals and principles of engineering surveying. Excellent results come from studying the *California Civil Surveying Reference Manual* by George M. Cole, Jr., PhD, PE, PLS; *1001 Solved Surveying Fundamentals Problems*, by Jan Van Sickle, PLS; and *California Civil Surveying Solved Problems*, by Peter R. Boniface, PhD, PLS, and Allan Y. Ng, MCSE, PLS; these books are published by PPI. This book is designed to complement the other publications.

California Civil Surveying Practice Exams acquaints you with the test plan adopted by the California Board for Professional Engineers, Land Surveyors, and Geologists. The intent of these practice exams is to measure your preparedness. With them, you can appraise your knowledge and skills before taking the official exam. Solutions are presented with explanations of the relevant key points and essential steps for solving problems.

Engineers who have taken the California Civil Engineering Surveying Exam frequently comment that there is not enough time to complete the questions. You must therefore improve your problem-solving speed by becoming proficient in the use of a calculator—particularly with respect to computations involving angles. Angles, bearings, and distances are fundamental to virtually all survey applications. Unfortunately, most calculators do not easily handle degrees, minutes, and seconds, and the default angular unit is decimal degrees. Make a point of learning the functions that enable rapid computation of trigonometric functions as well as addition, subtraction, and multiplication of angles directly in degrees, minutes, and seconds.

Pay special attention to the following topics, which are important to surveying and form a necessary core of knowledge: coordinate geometry (COGO) functions (such as inverse, side shot, bearing-bearing intersection, traverse, areas), differential leveling, horizontal and vertical curves, earthwork volumes, and datums.

Each examinee has a unique learning style and approach to understanding the exam subject material. Self-study may suit you better than formal instruction, or vice versa, and it is important to prepare with the method that best helps you learn. In addition, taking these practice exams in a simulated examination situation with constraints similar to those of the official exam can measure your level of readiness. Testing yourself using practice exams and practice problems will increase your proficiency in and knowledge of all surveying subject areas covered in the exam, as well as improve your chances of passing.

THE NATURE OF THE EXAM

The California Civil Engineering Surveying Exam tests the entry-level competency of a candidate to practice civil engineering within the profession's acceptable standards for public safety. Administered as a computer-based test (CBT), this exam is open book and is administered over 2.5 hours. The exam contains 55 multiple-choice problems derived from content areas as outlined in the board-adopted Engineering Surveying Test Plan. For each problem, you will be asked to select the best answer from four choices.

The point values of each exam question are printed in the official test booklet. Points are assigned depending on the significance, difficulty, and complexity of the question.

WHAT TO TAKE TO THE EXAM

Come to the exam prepared with the following materials. Be sure to set them out ahead of time and devise a convenient way to carry them into the examination room with you.

- government-issued photo ID
- reference materials (only one box)
- two of the following four measuring devices: ruler, protractor, architect scale, engineer scale
- two non-QWERTY calculators

CONTENT AREAS AND PROBLEM DISTRIBUTION

The California Civil Engineering Surveying Exam is a difficult test and requires thorough preparation in all areas, including multiple-choice test-taking techniques. Easy and difficult problems, with variable point values, are distributed throughout the exam. Besides contending with the nature and difficulty of the exam itself, many examinees spend too much time on difficult problems and leave insufficient time to answer the easy ones. You should avoid this. The following system can be beneficial to you during the exam.

step 1: Work on the easy problems first and record your answers.

step 2: Work on problems that require minimal calculations and record your answers.

step 3: When you get to a problem that looks "impossible" to answer, go ahead and guess. Mark your "guess" answer on your answer sheet and continue.

step 4: When you face a problem that seems difficult but solvable, you may need considerable time to search for relevant information in your books, references, or notes. Continue to the next question.

step 5: When you come to a question that is solvable but you know requires lengthy calculations or is time-consuming, leave it for later. For this question, you know exactly where to look for relevant information in your books, references, or notes.

Based on the number of exam problems and allotted time, on the average you should not spend more than 2.5 minutes per problem. Thus, a lengthy or "time-consuming" question is one that will take you more than 2.5 minutes answer. The intent of this test-taking strategy is to save you precious time.

After you have gone over the entire exam, which problems you have already answered and those that still require your attention will be clear. You should now go over the exam a second time, with the following approach in mind.

step 1: The best and most successful approach is to go back and tackle the problems that require lengthy calculations; the problems that require time to search for relevant information in books, references, and notes; and finally the problems that seemed impossible.

step 2: Recheck your work for careless mistakes.

step 3: Set aside the last few minutes of your exam period to fill in a guess for any unanswered problems. There is no penalty for guessing. Only problems answered correctly will be counted toward your score.

HOW TO USE THIS BOOK

California Civil Surveying Practice Exams is designed to be used in conjunction with surveying textbooks. For optimal results, take these practice exams only after you have reviewed a wide range of textbooks on the surveying field and grasped the subject matter in depth. Familiarize yourself with the content and topics of the exam, as outlined in the board-adopted test plan, and focus your study efforts on those areas. Finally, solve as many practice problems as possible and consider applying some multiple-choice test-taking strategies as you take the practice exams.

You will benefit by simulating the official exam constraints and conditions and giving yourself 2.5 hours to answer the problems in each pratice exam; try using a timer. If friends or associates are also preparing for the exam, you may want to get together for a group exam simulation. When it comes to scoring and going over the provided solutions, a group discussion will help you to understand the subject matter more thoroughly.

When taking a practice exam, follow these steps.

step 1: Use the answer sheet provided to record your answers. Do not look at the solutions until after you have completed the practice exam.

step 2: Set a timer for 2.5 hours, and begin working.

step 3: When time is up, stop working on the problems. If you finish early, recheck your work during the remaining time. A habit of rechecking will serve you well on the official exam.

step 4: Use the answer key provided to check your results and determine your score. This step will enable you to tell what areas you may need to review for a better grasp of the subject matter.

step 5: Appraise your performance. If your point total is more than 65% of the perfect score, it can be considered a passing score.

step 6: Study the provided solutions, and review your textbooks and references for those areas where you either answered incorrectly or guessed. If you answered some questions correctly but feel a need to understand their concepts more in depth, study those areas as well.

EXAM SCORING

For the California Civil Engineering Surveying Exam, the board-adopted test plan lists the content areas of the exam and their assigned scoring percentages. The percentages assigned to each content area are the approximate proportion of total test points; the test plan does not reveal the total test points in advance (as it varies from exam to exam). This makes it difficult to anticipate the exact number of problems for each content area of the exam.

The official exam is graded against a "cut score"—a predetermined minimum passing score that varies from exam to exam. Historically, if you score above 65% of the total examination point value, you have a good chance of passing.

On the official exam, after initial scoring, any problem that the board finds to be flawed may be deleted. In the event of deletion, the point value of the deleted problem becomes zero and the total number of points possible on that exam is adjusted accordingly.

You will face the official exam with a higher probability of success by going through the scoring process on your practice exam. The scoring process will give you an idea of how to overcome your weaknesses and pass the exam.

Nomenclature

A	area	in^2, ft^2, yd^2
A'	area	yd^2
b	base	ft
BS	backsight	ft
C	correction factor	ft
C	cost	$
C_f	earth curvature correction	ft
CE	cost of excavation	$
CO	cost of overhaul	$
dep	departure	ft
dx	departure difference	ft
dy	latitude difference	ft
D	degree of curvature	deg
D	depth	ft
D	distance	ft
elev	elevation	ft
FS	foresight	ft
g	grade	%
h	height	ft
HD	horizontal distance	ft
HI	height of instrument	ft
I	intersection or interior angle	deg
IFS	intermediate foresight	ft
k	constant	deg
lat	latitude	ft
L	length	ft
L	length of curve	ft
P_l	tension (pull) on tape	lbf
r	rate of change of grade	%/ft
R	radius	ft
R	radius of the earth	ft
R	range	ft
R_f	refraction correction	ft
S	section area	ft^2
SD	slope distance	ft
T	standard temperature of tape	°F
T	tangent distance	ft
T_l	temperature of tape	°F
V	volume	yd^3
w	unit weight of tape	lbf/ft
W	width	ft

x	distance	ft
x	tangent offset	ft
y	elevation	ft
y'	tangent offset	ft

SYMBOLS

α	angle	deg
α	azimuth	deg
α	coefficient of thermal expansion of steel	1/°F
β	angle	deg
γ	angle	deg
θ	vertical angle or central angle	deg
σ	standard deviation	–
Δ	central angle	deg

SUBSCRIPTS

BM	benchmark
BVC	beginning of vertical curve
c	correction
l	length
m	mean or midway
PVI	point of vertical intersection
s	sag
S	section
TP	test point

Practice Exam 1 Answer Sheet

1. Ⓐ Ⓑ Ⓒ Ⓓ
2. Ⓐ Ⓑ Ⓒ Ⓓ
3. Ⓐ Ⓑ Ⓒ Ⓓ
4. Ⓐ Ⓑ Ⓒ Ⓓ
5. Ⓐ Ⓑ Ⓒ Ⓓ
6. Ⓐ Ⓑ Ⓒ Ⓓ
7. Ⓐ Ⓑ Ⓒ Ⓓ
8. Ⓐ Ⓑ Ⓒ Ⓓ
9. Ⓐ Ⓑ Ⓒ Ⓓ
10. Ⓐ Ⓑ Ⓒ Ⓓ
11. Ⓐ Ⓑ Ⓒ Ⓓ
12. Ⓐ Ⓑ Ⓒ Ⓓ
13. Ⓐ Ⓑ Ⓒ Ⓓ
14. Ⓐ Ⓑ Ⓒ Ⓓ
15. Ⓐ Ⓑ Ⓒ Ⓓ
16. Ⓐ Ⓑ Ⓒ Ⓓ
17. Ⓐ Ⓑ Ⓒ Ⓓ
18. Ⓐ Ⓑ Ⓒ Ⓓ
19. Ⓐ Ⓑ Ⓒ Ⓓ

20. Ⓐ Ⓑ Ⓒ Ⓓ
21. Ⓐ Ⓑ Ⓒ Ⓓ
22. Ⓐ Ⓑ Ⓒ Ⓓ
23. Ⓐ Ⓑ Ⓒ Ⓓ
24. Ⓐ Ⓑ Ⓒ Ⓓ
25. Ⓐ Ⓑ Ⓒ Ⓓ
26. Ⓐ Ⓑ Ⓒ Ⓓ
27. Ⓐ Ⓑ Ⓒ Ⓓ
28. Ⓐ Ⓑ Ⓒ Ⓓ
29. Ⓐ Ⓑ Ⓒ Ⓓ
30. Ⓐ Ⓑ Ⓒ Ⓓ
31. Ⓐ Ⓑ Ⓒ Ⓓ
32. Ⓐ Ⓑ Ⓒ Ⓓ
33. Ⓐ Ⓑ Ⓒ Ⓓ
34. Ⓐ Ⓑ Ⓒ Ⓓ
35. Ⓐ Ⓑ Ⓒ Ⓓ
36. Ⓐ Ⓑ Ⓒ Ⓓ
37. Ⓐ Ⓑ Ⓒ Ⓓ
38. Ⓐ Ⓑ Ⓒ Ⓓ

39. Ⓐ Ⓑ Ⓒ Ⓓ
40. Ⓐ Ⓑ Ⓒ Ⓓ
41. Ⓐ Ⓑ Ⓒ Ⓓ
42. Ⓐ Ⓑ Ⓒ Ⓓ
43. Ⓐ Ⓑ Ⓒ Ⓓ
44. Ⓐ Ⓑ Ⓒ Ⓓ
45. Ⓐ Ⓑ Ⓒ Ⓓ
46. Ⓐ Ⓑ Ⓒ Ⓓ
47. Ⓐ Ⓑ Ⓒ Ⓓ
48. Ⓐ Ⓑ Ⓒ Ⓓ
49. Ⓐ Ⓑ Ⓒ Ⓓ
50. Ⓐ Ⓑ Ⓒ Ⓓ
51. Ⓐ Ⓑ Ⓒ Ⓓ
52. Ⓐ Ⓑ Ⓒ Ⓓ
53. Ⓐ Ⓑ Ⓒ Ⓓ
54. Ⓐ Ⓑ Ⓒ Ⓓ
55. Ⓐ Ⓑ Ⓒ Ⓓ

Practice Exam 1

43/55 ≈ 78%.

1. Control surveys for photogrammetric mapping should position the control points according to   A

   (A) the number and positions of photos and the contour interval

   (B) the nature of the terrain and the location of the high points

   (C) visibility of adjacent traverse points or global positioning system (GPS) satellites

   (D) site accessibility

2. A closed traverse is preferred to a radial traverse because   A

   (A) points on a radial traverse are not checked

   (B) a closed traverse is more accurate

   (C) a closed traverse involves fewer observations

   (D) points are more easily visible on a closed traverse

3. A benchmark is generally defined as a   A

   (A) permanently marked point with a known elevation

   (B) permanently marked point with a known latitude and longitude

   (C) continuously recording global positioning system (GPS) point

   (D) point that defines mean sea level in a tidal area of water

4. A digital alternative to a contour map is called a   B

   (A) digital terrain model

   (B) digital elevation model

   (C) TIN file

   (D) triangulated network

5. At a river crossing where it is impossible to equalize the backsight and foresight, which of the following leveling methods should be used?   D

   (A) profile

   (B) barometric

   (C) trigonometric

   (D) reciprocal

6. On a closed differential leveling loop, the sum of the backsights minus the sum of the foresights should equal   A

   (A) zero

   (B) the difference between the highest and lowest points on the loop

   (C) an arbitrary number depending on the steepness of the terrain

   (D) twice the value of the difference between the highest and lowest points of the loop

7. Consider a horizontal distance, PR, of 50.00 ft. The profile leveling field data for PQ is shown in the table.   A

point	BS	FS
P	4.67	–
Q	–	5.78
R	–	7.22

The grade of course PR is most nearly

   (A) −5.1%

   (B) −2.9%

   (C) 2.9%

   (D) 5.1%

8. Differential leveling measurements for a site are given in the table shown.

point	BS (ft)	FS (ft)	elevation (ft)
A	3.34	–	1000.00
B	5.00	2.89	–
C	4.78	3.04	–
D	–	1.11	1006.18

The height of instrument (HI) at the setup between points B and C is most nearly

(A) 994.55 ft

(B) 1000.45 ft

(C) 1002.41 ft

(D) 1005.45 ft

9. A highway centerline is defined by station L and station M.

$$\text{sta L} = 15 + 34.86 \text{ sta}$$
$$\text{sta M} = 17 + 09.08 \text{ sta}$$

A level is set up between station L and station M. The backsight to L is 3.78 ft. The foresight to M is 8.10 ft. Most nearly, the grade line of LM is

(A) 0.10%

(B) 2.5%

(C) 4.6%

(D) 6.8%

10. A 12 ft leveling rod is held next to a roof such that the top of the rod (12.00 ft mark) is level with the edge of the roof. A level is set up in between the edge of the roof and the temporary benchmark. A backsight is then taken to the edge of the roof and has a reading of 4.89 ft. The instrument is turned and a temporary benchmark is sighted; the rod reading is 7.90 ft. The height difference between the benchmark and the edge of the roof is most nearly

(A) 0.79 ft

(B) 1.79 ft

(C) 14.01 ft

(D) 15.01 ft

11. An instrument set up over point A obtains a rod reading (RR) from a rod set up at point B. The elevation at point A is 937.4 ft, the elevation at point B is 932.1 ft, and the rod reading is 9.1 ft. The height of the instrument (HI) is most nearly

(A) 2.7 ft

(B) 3.8 ft

(C) 4.6 ft

(D) 5.3 ft

12. The rod used for leveling of the highest precision is made of

(A) well-seasoned wood

(B) an Invar® strip

(C) stainless steel

(D) an aluminum alloy

13. The most common elevation datum for modern surveys is

(A) NAD 27

(B) NAD 83

(C) NGVD 29

(D) NAVD 88

14. Before computing a traverse on NAD 83, measured distances must be corrected for

(A) grid scale factor

(B) sea-level scale factor

(C) neither (A) nor (B)

(D) both (A) and (B)

15. A benchmark set for a current engineering project in the continental United States most likely refers to the

(A) North American Vertical Datum of 1988 (NAVD 88)

(B) American Samoa Vertical Datum of 2002 (ASVD02)

(C) Northern Marianas Vertical Datum of 2003 (NMVD03)

(D) Virgin Islands Vertical Datum of 2009 (VIVD09)

16. The horizontal control datum used for the United States is   C

(A) NAD 27
(B) NAD 29
(C) NAD 83
(D) NAD 88

17. A photo image in which every pixel is in its correct map position is known as a   C

(A) photomap
(B) digital image
(C) digital orthophoto·
(D) raster map

18. On a contour map of an area compiled from aerial photos using a stereoplotter, the contours are usually generated by   B

(A) directly drawing the contours by stereo measurement
(B) computer interpolation from a digital elevation model
(C) computer interpolation from profiles
(D) automatically generating contours by image-matching

19. If the overlap on a pair of aerial 9 in by 9 in framed photos is 60% with a photo scale of 1 in:500 ft, most nearly, the area covered by the two overlapping photos is   B

(A) 0.20 mi²
(B) 0.44 mi²
(C) 0.60 mi²
(D) 0.75 mi²

20. The minimum ground control required to map from an overlapping pair of aerial photos is   C

(A) a plan point and a height point in each corner of the overlap
(B) three plan points and three height points on the overlap that do not fall in a straight line
(C) three height points that do not fall in a straight line, and two plan points on the overlap
(D) two plan points and two height points, both in opposite corners of the overlap

21. A table of data pertains to an open four-course traverse.

side	distance (ft)	azimuth	point	coordinate y	coordinate x
			A	560.00	770.00
AB	394.59	81°14′54″			
			B		
BC	292.75	172°08′50″			
			C		
CD	332.87	212°44′26″			
			D		
DE	323.88	351°07′42″			
			E		

The (y, x) coordinates of point E are most nearly

(A) (370.05, 970.02)
(B) (687.54, 644.10)
(C) (749.95, 569.98)
(D) (760.02, 580.05)

22. For the points shown,

(y, x) coordinates of L = (540.55, 879.01)
(y, x) coordinates of M = (638.78, 925.43)
bearing MP = due west
bearing LP = N 68°35′00″ W

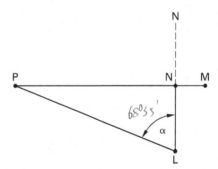

The (y, x) coordinates of point P are most nearly

(A) (638.78, 628.57)
(B) (628.57, 638.78)
(C) (638.78, 840.48)
(D) (638.78, 1129.45)

23. A region and its parameters are shown.

(y, x) coordinate of A   (1675.24, 2546.09)
(y, x) coordinate of B   (1294.39, 2870.06)
azimuth AP   95°34'20"
azimuth BP   48°00'00"

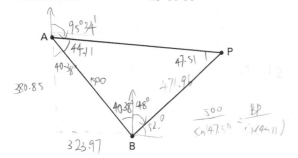

The (y, x) coordinates of point P are most nearly
- (A) (−1609.49, −3220.02)
- (B) (1474.36, 3892.11)
- (C) (1598.34, 3564.19)
- (D) (1609.49, 3220.02)

24. A line PQ is measured and recorded as a slope distance (SD) of 257.56 ft with a slope of 4°00'00". If the actual slope of the line is 3°00'00", the error introduced in horizontal distance as a result of the incorrect slope is most nearly
- (A) 0.16 ft
- (B) 0.28 ft
- (C) 0.96 ft
- (D) 1.28 ft

25. A large parcel of land is subdivided such that the area of parcel 1 is 6.00 ac, as shown.

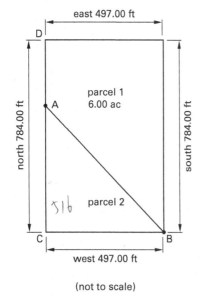

(not to scale)

Most nearly, the length of boundary AB is
- (A) 690 ft
- (B) 720 ft
- (C) 740 ft
- (D) 750 ft

26. The owners of lot 1 and lot 2 have agreed to an adjustment of the lot line. The current property line and the proposed line to the west are shown.

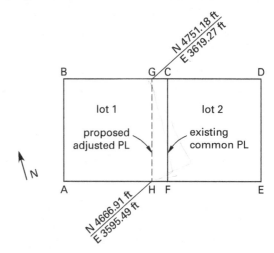

Given the coordinates provided, the length of the adjusted property line, GH, is most nearly
- (A) 81.9 ft
- (B) 85.3 ft
- (C) 87.6 ft
- (D) 91.7 ft

27. Sides a, b, and c of the triangle shown were measured as 330.56 ft, 210.90 ft, and 380.02 ft, respectively.

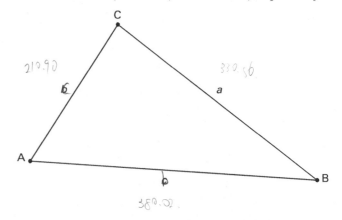

Angle B is most nearly

(A) 33°26′05″   C

(B) 33°27′33″

(C) 33°37′23″

(D) 34°02′30″

28. A roadway is shown.

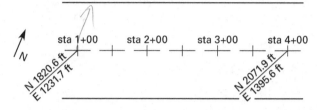

Most nearly, the bearing of the roadway centerline shown is

(A) N 33°07′ E   A

(B) S 33°07′ W

(C) N 56°53′ E

(D) S 56°53′ W

29. A pair of triangles is shown.

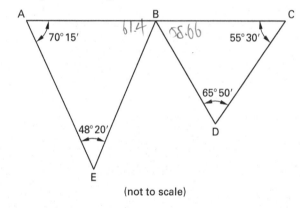

(not to scale)

Line AB and line BC have the same bearing. The value of angle EBD is most nearly

(A) 30°05′   C

(B) 58°40′

(C) 59°55′

(D) 61°25′

30. On the 2000 ft radius curve shown, the tangent distance is 278.346 ft, and the tangent offset is 19.464 ft.

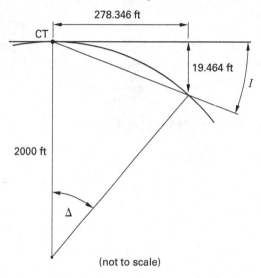

(not to scale)

Determine the central angle, Δ.

(A) 2°   D

(B) 4°

(C) 6°

(D) 8°

31. A horizontal curve is shown.

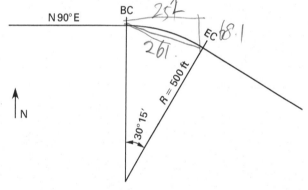

The (x, y) coordinates of point BC are (100.4 ft, 92.1 ft). Most nearly, the (x, y) coordinates of point EC are

(A) (151.5 ft, 24.0 ft)   C

(B) (151.5 ft, 160.2 ft)

(C) (352.3 ft, 24.0 ft)

(D) (352.3 ft, 160.2 ft)

32. A curve has the characteristics shown.

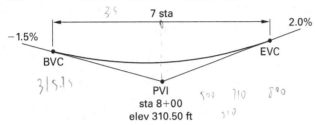

The elevation of sta 7+10 on the curve is most nearly

(A)   311.77 ft

(B)   312.09 ft

(C)   313.54 ft

(D)   321.34 ft

$315.75 - 0.015 \times 210ft + 0.25210ft^2$
0.25×2.1^2

33. A symmetric vertical curve is shown.

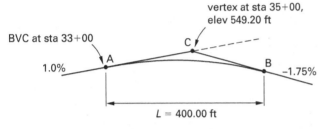

At what station does the curve crest?

(A)   sta 33+93.71

(B)   sta 34+35.40

(C)   sta 34+45.45

(D)   sta 35+11.00

$\frac{-1.75-1}{4}$

$-1\frac{1}{16}\ \%/sta$

34. A reverse curve has the characteristics shown.

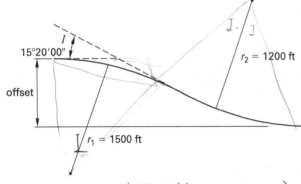

(not to scale)

$x = (r_1 + r_2)(1 - \cos I)$

$c\ \sin \frac{d}{2}$

The offset is most nearly

(A)   10.68 ft

(B)   79.33 ft

(C)   96.11 ft

(D)   713.97 ft

35. The radii and central angles of a compound curve are 2000 ft, 15°05'00" and 2450 ft, 11°45'00". The total length of the compound curve is most nearly

(A)   1014.95 ft

(B)   1028.94 ft

(C)   1047.40 ft

(D)   1055.12 ft

36. Refer to the illustration shown.

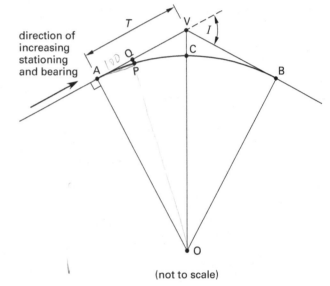

(not to scale)

Point P lies on a 4000 ft radius curve. Point Q lies on the tangent to the curve, and line PQ is perpendicular to the tangent. The distance AQ is 100 ft. Most nearly, what is the distance PQ?

(A)   1.3 ft

(B)   2.5 ft

(C)   3.0 ft

(D)   9.5 ft

$\frac{\sin(d)}{2}$

$TD = 2R\ \sin\frac{d}{2} \cdot \cos\frac{d}{2}$

$100 = 8000 \cdot \sin\frac{d}{2} \cdot \cos\frac{d}{2}$

$TO = 0.2R \cdot \sin\frac{d}{2} \cdot \sin\frac{d}{2}$

$TO = 8000 \cdot \sin\frac{d}{2} \cdot \sin\frac{d}{2}$

37. During an improvement project, a construction staking survey is typically performed   B

(A) prior to design of the project

(B) prior to the beginning of construction

(C) during project construction

(D) after construction is complete

38. Refer to the illustration shown.

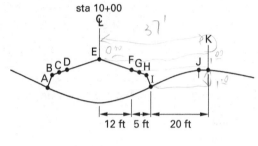

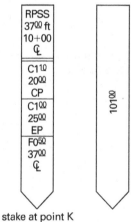

sta 10+00

stake at point K
(front)   (back)

(not to scale)

EP: Edge of Pavement

CP: catch point of a slope

The second entry under the double strike line on the coded stake at point K indicates that the

D

(A) excavation at point F is 1.0 ft

(B) embankment at point F is 1.0 ft

(C) difference in elevation between point G and point F is 1.0 ft

(D) difference in elevation between the finished grade at point F is 1.0 ft below the elevation of the stake set at point K

39. Refer to the illustration shown.

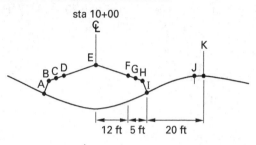

sta 10+00

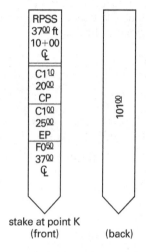

stake at point K
(front)   (back)

(not to scale)

The first entry under the double strike line on the coded stake at point K indicates that the   D

(A) excavation required at point H is 1.1 ft

(B) finished grade elevation at point H is designed to be 1.1 ft below the elevation of the stake set at point K

(C) embankment required at point H is 1.1 ft

(D) difference in elevation between point I and point K is 1.1 ft

40. Control surveys that are used for aerial mapping projects are usually based on   B

(A) total station

(B) global positioning system

(C) triangulation

(D) stadia

41. A line AB is measured with steel tape three times.

1st measurement: 597.410 ft at a temperature of 83°F

2nd measurement: 597.320 ft at a temperature of 98°F

3rd measurement: 597.500 ft at a temperature of 49°F

The standard temperature of the tape is 68°F, and the coefficient of expansion of steel is 6.45×10^{-6}/°F. The length of line AB is most nearly *C*

(A) 597.331 ft

(B) 597.384 ft

(C) 597.444 ft

(D) 597.492 ft

42. A straight line 8.6 km in length is trig-leveled. Most nearly, what is the combined effect of earth curvature and refraction on this line?

A

(A) 16.40 ft

(B) 19.02 ft   *C−R*

(C) 20.07 ft

(D) 21.65 ft

43. A closed traverse totaling 3250 ft in length closes onto a point with known (x, y) coordinates (2459.16 ft, 5211.90 ft). The computed coordinates of this point after the computation of the last traverse course are (2459.38 ft, 5211.71 ft). Most nearly, the accuracy of the traverse is

B

(A) 1:8000

(B) 1:11,000

(C) 1:12,000

(D) 1:15,000

44. The observed angles of a four-course closed traverse are

C

angle A	88°47′30″
angle B	110°15′50″
angle C	79°00′10″
angle D	81°55′50″

The balanced angle D is most nearly

(A) 81°55′10″

(B) 81°55′40″

(C) 81°56′00″

(D) 81°56′40″

DON'T FORGET TO DIVIDE ERROR n̄

45. Leveling field notes for benchmark (BM), backsight (BS), and foresight (FS) are given.

station	BS (ft)	HI (ft)	FS (ft)	elevation (ft)
BM1	4.20			431.20 (known)
TP1	5.34		3.98	
TP2	8.45		8.30	
BM2			7.62	429.37 (known)

Most nearly, the vertical error of closure is

B

(A) −0.48 ft

(B) −0.08 ft   *measured − known*

(C) 0.08 ft

(D) 0.48 ft

46. The general methods of keeping survey field notes are

B

I. writing a description of the survey work performed

II. taking representative photographs of all survey operations

III. creating a detailed sketch of the survey with numeric values shown

IV. using the survey instrument's automatic storage of field measurements

V. tabulating the survey's numeric values

(A) I, II, and IV

(B) I, III, and V

(C) II, IV, and V

(D) III, IV, and V

47. What is most nearly the area of a sector with a central angle of 47° within a curve that has a radius of 1100 ft?

D

(A) 10.3 ac

(B) 10.8 ac $\dfrac{A}{\pi R^2} = \dfrac{I}{360°}$

(C) 11.0 ac

(D) 11.4 ac

48. A cross section is shown.

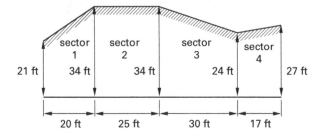

Most nearly, the area of the cross section shown is

(A) 100 yd²

(B) 160 yd²

(C) 240 yd²

(D) 300 yd²   285.33

49. A roadway cross section is shown.

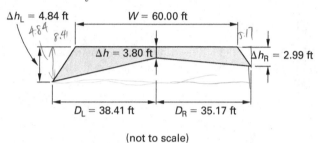

(not to scale)

Most nearly, the area of the given roadway cross section is

(A) 190 ft²   B

(B) 260 ft² $A = \Delta h\left(\frac{D_L + D_R}{2}\right) + W\left(\frac{\Delta h_L + \Delta h_R}{4}\right)$

(C) 300 ft²

(D) 370 ft²

50. Dirt is excavated from a borrow pit 80 ft by 60 ft by 5 ft deep and moved 800 ft away. The cost of excavation is \$2.00/yd³. The first five stations are free. The cost of overhaul for each station over five stations is \$0.40/yd³. What is most nearly the total cost of the move?   A   88888

(A) \$2850

(B) \$3000

(C) \$3500

(D) \$4650

51. For the computation of an earthwork quantity between two cross sections using the prismoidal formula, the middle section dimensions are obtained by

(A) measurement in the field   B

(B) averaging the dimensions of the two end cross sections

(C) averaging the dimensions of the two end cross sections using a weighted average

(D) either area of the end cross sections

52. A topography map is shown.   2.5

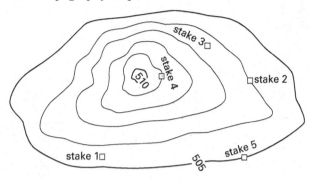

Construction stakes are placed on existing terrain. The horizontal distance between stake 3 and stake 4 is 50 ft. What is most nearly the grade between these points?

(A) 3%   C

(B) 4%

(C) 5%

(D) 6%

53. A survey data table is shown.   A

station	EG elevation (ft)
1+00	500.0
1+25	515.3
1+50	530.4
1+75	540.7
2+00	539.2
2+25	528.9
2+50	514.0
2+75	501.6
3+00	498.2

Which ground profile best matches the data provided?

(A) ✓

sta 1+00   sta 2+00   sta 3+00

(B)

sta 1+00   sta 2+00   sta 3+00

(C)

sta 1+00   sta 2+00   sta 3+00

(D)

sta 1+00   sta 2+00   sta 3+00

54. An aerial photograph is taken using a camera with a focal length of 6 in. The plane flies at an altitude of 4000 ft above mean sea level, and the mean ground elevation is 400 ft above mean sea level. The scale of the photograph is most nearly

(A)   1 in:10 ft

(B)   1 in:50 ft

(C)   1 in:600 ft

(D)   1 ft:10 ft

$S = \dfrac{f}{H}$

$H = $ altitude $-$ mean ground elev

55. A soccer field 120 yd in length measures 9.144 mm on an aerial photo. The scale of the aerial photo is most nearly

(A)   1:4000

(B)   1:10,000

(C)   1:12,000

(D)   1:110,000

Practice Exam 2 Answer Sheet

56. (A) (B) (C) (D)
57. (A) (B) (C) (D)
58. (A) (B) (C) (D)
59. (A) (B) (C) (D)
60. (A) (B) (C) (D)
61. (A) (B) (C) (D)
62. (A) (B) (C) (D)
63. (A) (B) (C) (D)
64. (A) (B) (C) (D)
65. (A) (B) (C) (D)
66. (A) (B) (C) (D)
67. (A) (B) (C) (D)
68. (A) (B) (C) (D)
69. (A) (B) (C) (D)
70. (A) (B) (C) (D)
71. (A) (B) (C) (D)
72. (A) (B) (C) (D)
73. (A) (B) (C) (D)
74. (A) (B) (C) (D)

75. (A) (B) (C) (D)
76. (A) (B) (C) (D)
77. (A) (B) (C) (D)
78. (A) (B) (C) (D)
79. (A) (B) (C) (D)
80. (A) (B) (C) (D)
81. (A) (B) (C) (D)
82. (A) (B) (C) (D)
83. (A) (B) (C) (D)
84. (A) (B) (C) (D)
85. (A) (B) (C) (D)
86. (A) (B) (C) (D)
87. (A) (B) (C) (D)
88. (A) (B) (C) (D)
89. (A) (B) (C) (D)
90. (A) (B) (C) (D)
91. (A) (B) (C) (D)
92. (A) (B) (C) (D)
93. (A) (B) (C) (D)

94. (A) (B) (C) (D)
95. (A) (B) (C) (D)
96. (A) (B) (C) (D)
97. (A) (B) (C) (D)
98. (A) (B) (C) (D)
99. (A) (B) (C) (D)
100. (A) (B) (C) (D)
101. (A) (B) (C) (D)
102. (A) (B) (C) (D)
103. (A) (B) (C) (D)
104. (A) (B) (C) (D)
105. (A) (B) (C) (D)
106. (A) (B) (C) (D)
107. (A) (B) (C) (D)
108. (A) (B) (C) (D)
109. (A) (B) (C) (D)
110. (A) (B) (C) (D)

Practice Exam 2

56. A control survey   *C*

(A) uses a level to control the elevations of cut/fill on an alignment

(B) finds radial measurements from a single base station, controlling the setting out of engineering structures

(C) sets out a network of points covering a site, for the future control of additional survey measurements

(D) inspects completed earthworks that act as a control on the final earthwork quantities

57. The geoid is the   *C*

(A) geodetic reference surface (a rotated ellipse) that is the basis of the state-plane coordinate system

(B) deviation of mean sea level from the reference spheroid

(C) mean sea-level surface covering the entire globe

(D) actual surface of the earth both above and below sea level

58. The purpose of a control survey for an engineering project is to   *D*

(A) locate existing terrain features within the project limits

(B) layout proposed construction improvements on the project site

(C) identify the alignment of proposed features, such as walls, pipes, and roadway centerlines

(D) establish the horizontal and vertical reference positions of the project

59. Topographic surveys can be used to establish

  I. terrain contours

  II. spot elevations

  III. the location of existing fence lines

  IV. the limits of construction

                      C

(A) I only

(B) I and II

(C) I, II, and III

(D) I, II, III, and IV

60. An 18 in length of rebar is driven into the ground until it touches the top of an underground pipe. A backsight to a benchmark with an elevation of 390.66 ft is taken with a tilting level. The backsight has a rod reading of 7.56 ft. A foresight is taken to a rod resting on top of the rebar. The rod reading is 1.01 ft. The elevation of the top of the pipe is most nearly

(A) 382.61 ft

(B) 385.95 ft

(C) 395.71 ft

(D) 398.71 ft

61. A backsight to a benchmark at a point A reads 2.55 ft. The elevation of point A is 2800.20 ft above mean sea level. A surveyor is required to place a mark on a lath, which would have an elevation of 2798.11 ft. The rodman holds the rod next to the lath. To set the base of the rod to this elevation, the rod reading (RR) would be most nearly

                               D

(A) 0.56 ft

(B) 0.76 ft

(C) 4.46 ft

(D) 4.64 ft

62. Backsight (BS), foresight (FS), and elevation data for two points, P and Q, are given in the table.

point	BS (ft)	FS (ft)	elevation (ft)
P	4.68	–	1000.00
	3.99	2.12	–
	2.07	1.63	–
	5.55	1.64	–
Q	–	4.71	–

The elevation of point Q is most nearly

(A)   993.81 ft   C

(B)   1003.98 ft

(C)   1006.19 ft

(D)   1015.61 ft

63. At a setup during profile leveling, three crosshair readings are recorded at points A and B.

point	A (ft)	B (ft)
top	5.15	6.30
mid	5.00	5.92
bottom	4.85	5.54

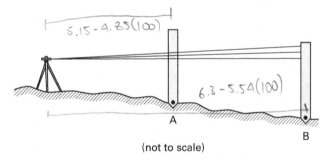

(not to scale)

Assume setup points, A and B, are in a straight line. If $k = 100$ and $C = 0$, the grade of line AB is most nearly

(A)   –2%   A

(B)   –1%

(C)   1%

(D)   2%

64. During a profile leveling run, the backsight to point A, which has an elevation of 500 ft, is 4.77 ft. The foresights to point B and point C are 3.65 ft and 5.89 ft,

respectively. Most nearly, the elevation of point B is higher (+) or lower (−) than point C by

(A)   −3.24 ft

(B)   −2.24 ft

(C)   2.24 ft

(D)   3.24 ft

65. A backsight of 1.67 ft is read to a benchmark with a known elevation of 378.54 ft. The foresight to the rod resting on an exposed horizontal pipe reads 8.06 ft. The pipe has an inside diameter of 5 ft 4 in, and the thickness of the pipe casing is 1 in. The elevation of the center of the pipe is most nearly

(A)   369.40 ft   A

(B)   377.57 ft

(C)   379.51 ft

(D)   390.35 ft

66. The table shown gives observations made in a profile leveling.

Intermediate Foresight

point	BS (ft)	HI (ft)	FS (ft)	IFS (ft)	elevation (ft)
A	4.87	–	–	–	100.00
B	–	–	–	6.87	–
C	–	–	–	5.91	–
D	–	–	–	4.77	–

The elevation of point D is most nearly

(A)   87.32 ft   C

(B)   99.90 ft   100 + 4.87 − 4.77

(C)   100.10 ft

(D)   112.68 ft

67. An instrument at point A is backsighting benchmark 1. The elevation at point A is 1059.4 ft, and the elevation at benchmark 1 (BM1) is 1056.6 ft. The rod reading (RR) at BM1 is 6.2 ft. The instrument obtains a foresight reading at point B. The RR at point B is 3.8 ft. Most nearly, the elevation at point B is

(A)   1059.0 ft   A

(B)   1059.8 ft

(C)   1064.6 ft

(D)   1066.6 ft

68. Which of the following datum are originally legislated with meter? *B*

(A) North American Datum (NAD) of 1927

(B) North American Datum (NAD) of 1983

(C) National Geodetic Vertical Datum (NGVD) of 1929

(D) U.S. Standard Datum (USSD) of 1988

69. The national leveling network that was adjusted to fit 26 mean sea-level stations is

(A) NAVD 88   — one point   *B*

(B) NGVD 29   — 26 points

(C) NAD 83

(D) NAD 27

70. The national vertical datum that is not a sea-level datum and was held fixed to a tidal benchmark in Quebec is known as   *D*

(A) NAD 27   — horiz

(B) CAN 91   — does not exist

(C) NGVD 29   — 26 points

(D) NAVD 88

71. Prior to 1991, the datum for height above sea level was   *C*

(A) NAD 27

(B) NAD 83

(C) NGVD 29   used before 1991

(D) NAVD 88   used after 1991

72. The scale of a stereo overlap formed from two aerial photos is determined from surveyed ground control. When redundant control is provided in the form of a third planimetric control point, the scale is determined from   *B*

(A) the longest side between two control points

(B) a least-squares solution using all three control points

(C) the mean computed from three distances between control points

(D) a weighted mean computed from three distances between control points

73. Full control on an aerial stereo overlap requires   *B*

(A) two points with a known plan position and elevation

(B) three elevation points (one in each corner) and two plan points

(C) four points (one in each corner) with a known plan position and elevation

(D) six points (one in each corner and one at each photo center) with a known plan position and elevation

74. The distance from point A to a point B is 130.65 ft and the bearing is N 18°56′00″ W. The (x, y) coordinates of point A are (310.00 ft, 275.00 ft). The (x, y) coordinates of point B are most nearly

(A) (258.54 ft, 261.99 ft)

(B) (267.61 ft, 398.58 ft)

(C) (352.39 ft, 398.58 ft)

(D) (433.58 ft, 232.61 ft)

75. Points are shown on the illustration.

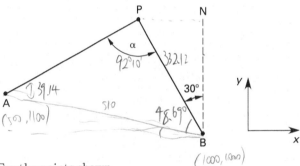

For the points shown,

$$\text{bearing BP} = \text{N } 30°00′00″ \text{ W}$$
$$\text{angle } \alpha = 92°10′00″$$
$$(x, y) \text{ coordinates of point A} = (500 \text{ ft}, 1100 \text{ ft})$$
$$(x, y) \text{ coordinates of point B} = (1000 \text{ ft}, 1000 \text{ ft})$$

The (x, y) coordinates of point P are most nearly   *D*

(A) (−873.49 ft, −1278.96 ft)

(B) (322.05 ft, 1006.05 ft)

(C) (838.72 ft, 1279.34 ft)

(D) (838.94 ft, 1278.96 ft)

76. For the circular arc shown, the (x, y) coordinates of the center, C, are (5000 ft, 5000 ft).

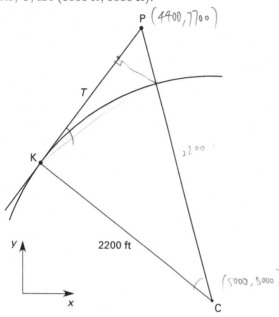

If the (x, y) coordinates of point P are (4400 ft, 7700 ft), the tangent distance, T, is most nearly

(A) 1587.43 ft

(B) 1676.30 ft

(C) 2887.30 ft

(D) 3177.46 ft

77. A 300 ft steel tape is used to place two stakes: one at the 0 ft mark and one at the 300 ft mark. The coefficient of thermal expansion of steel, α, is $6.45 \times 10^{-6}/°F$. The air temperature is 93°F. If the standard temperature of the steel tape used is 68°F, the true distance between the stakes is most nearly

(A) 299.95 ft

(B) 299.98 ft

(C) 300.02 ft

(D) 300.05 ft

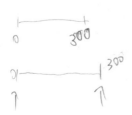

78. A sloped excavation site is shown.

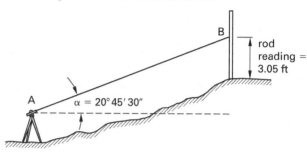

height of instrument at A	4.52 ft
elevation of A	497.26 ft
slope distance (SD) (measured along line of sight)	187.44 ft

The elevation of stake B is most nearly

(A) 565.16 ft

(B) 569.78 ft

(C) 571.26 ft

(D) 575.89 ft

79. The (x, y) coordinates of point A and point B are (101.56 ft, 556.23 ft) and (637.89 ft, 15.33 ft), respectively. The radius of the circular arc (center at B) is 700.00 ft.

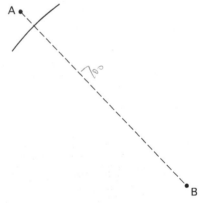

The shortest distance from A to the arc is most nearly

(A) 7.86 ft

(B) 61.7 ft

(C) 87.9 ft

(D) 762 ft

80. Radial traverse PQ, PR, PS, PT has observed angles α, β, and θ, as shown.

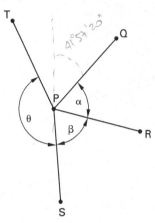

The bearing of PQ is N 41°57′20″ E, and the observed angles are as follows.

$$\alpha = 62°10'10''$$
$$\beta = 71°44'30''$$
$$\theta = 158°32'40''$$

The bearing of PT is most nearly B

(A) N 25°35′20″ E

(B) N 25°35′20″ W

(C) N 35°35′20″ W

(D) N 64°24′40″ W

81. An open traverse is shown.

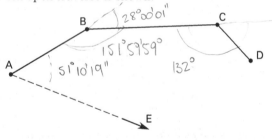

azimuth AE	111°47′36″
interior angle EAB	51°10′19″
deflection angle at B	28°00′01″
interior angle BCD	132°00′00″

$60°37°17°$

The azimuth of line CD is most nearly B

(A) 80°37′16″

(B) 136°37′18″

(C) 238°57′56″

(D) 260°37′16″

82. A bearing line is shown.

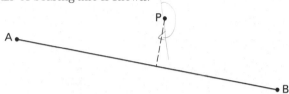

Line BA is N 79°34′50″ W. The azimuth of the shortest line from point P to line AB is most nearly

(A) 10°25′10″ C

(B) 169°34′50″

(C) 190°25′10″

(D) 349°34′50″

83. The layout control point for a horizontal curve is set at the BC, located at sta 143+52. Construction stakes are set along the curve at stations every 50 ft. The radius of the curve is 500 ft and the central angle is 40°. Most nearly, what is the straight (chord) distance from the layout control point to the sixth stake?

(A) 245.5 ft

(B) 293.6 ft

(C) 298.0 ft

(D) 342.0 ft

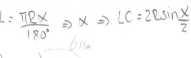

84. On a horizontal curve with a radius of 2400 ft and an intersection angle of 15°30′00″, the distance from the point of intersection (PI) to the nearest point on the curve is most nearly

(A) 20.71 ft

(B) 21.92 ft

(C) 22.12 ft

(D) 22.21 ft

85. A horizontal curve has a degree of curvature that measures 2°30′00″. The deflection angle of the curve at the point of intersection, I, is 12°00′00″. The stationing at the beginning of curve is sta 21+03.90. If points are staked every 100 ft on the curve, the deflection angle at the end of curve from the tangent to the second point on the curve will be most nearly

(A) 3°30′59″

(B) 3°32′56″

(C) 4°47′56″

(D) 7°05′51″

$L_{TOTAL} = \dfrac{I}{D}(100ft)$

$= 480$

B

86. A curve has the characteristics shown.

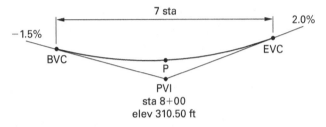

The elevation of the lowest point, P, on the curve is most nearly

(A) 313.50 ft   A

(B) 322.50 ft

(C) 323.67 ft

(D) 325.01 ft

87. A curve is shown.

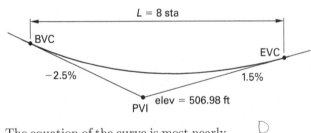

The equation of the curve is most nearly   D

(A) $y = 516.98 \text{ ft} - 2.5x + 0.5x^2$

(B) $y = 496.98 \text{ ft} - 1.5x + 0.25x^2$

(C) $y = 496.98 \text{ ft} + 2.5x + 0.5x^2$

(D) $y = 516.98 \text{ ft} - 2.5x + 0.25x^2$

88. An existing highway is detoured due to an obstruction, as shown. A radius of 1000 ft must be held for each curve in a reverse curve.

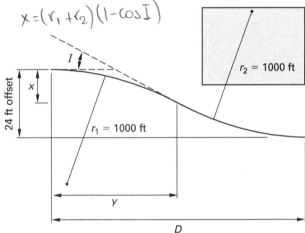

$x = (r_1 + r_2)(1 - \cos I)$

(not to scale)

219.08

12.58

The distance required to develop a 24 ft offset is most nearly   C

(A) 77.23 ft

(B) 154.45 ft $2R \sin \frac{d}{2}$ $\cdot \sin \frac{d}{2}$

(C) 308.90 ft

(D) 439.09 ft 15492

89. A compound curve is shown.

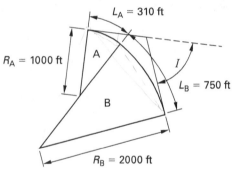

The total intersection angle, I, for the compound curve shown is most nearly

(A) 39°09′57″   C

(B) 39°10′26″

(C) 39°14′51″

(D) 39°25′16″

90. A benchmark, BMc, has an elevation of 763.60 ft, and the horizontal distance from BMb to BMc is 200.00 ft. Backsights and foresights are given in the table.

point	BS (ft)	FS (ft)	elevation (ft)
BMa	4.67	–	783.67
	3.00	5.59	–
	4.26	6.51	–
	3.22	5.74	–
BMb	–	7.38	–
BMc	–	–	763.60

The ground between BMb and BMc has a uniform grade. The grade of the line from BMb to BMc is most nearly

A

(A) −5%

(B) −2%

(C) 2%

(D) 5%

91. A curve is shown.

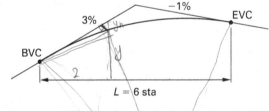

The tangent offset to the vertical curve at $x = 2.00$ sta from the beginning of the vertical curve (BVC) is most nearly

(A) −2.00 ft

(B) −1.32 ft

(C) 1.32 ft

(D) 2.00 ft

$$y = \left(\frac{r}{2}\right)x^2$$
$$= \frac{-0.66\,(2)^2}{2} =$$

B

92. In construction staking, a 4 ft stake showing information relating to another stake is called a(n)

(A) witness stake   *D*

(B) hub

(C) offset stake

(D) lath

93. Refer to the illustration shown.

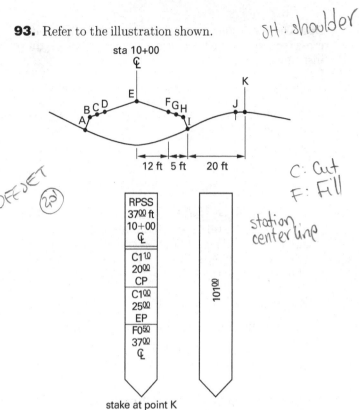

SH: shoulder

OFFSET (2)

C: Cut
F: Fill

station
centerline

The code on the back of the stake at point K indicates that the   *A*

(A) elevation of the top of the stake at point J is 101.00 ft

(B) offset distance from point K to point E is 101.00 ft

(C) design finished grade elevation of point E is 101.00 ft

(D) offset distance from point A to point I is 101.00 ft

94. The term RPSS on a construction stake stands for

B

(A) reference point side stake

(B) reference point slope stake

(C) right point side slope

(D) rough point side stake

95. The elevations of a terrain are measured along a cross section on a new highway alignment. The height of instrument (HI) is

B

(A) measured with a tape

(B) determined from the backsight to a point with known elevation

(C) determined from a foresight

(D) not required

96. In the "two-peg test" in leveling, backsight and fore-sight readings are taken at a setup midway between two points P and Q. Then, both points are sighted from an instrument set up outside of line PQ, and the results are used to determine the   C

(A) errors on the rods

(B) reading error of the surveyor

(C) mislevelment of the line of sight of the level

(D) accuracy of the circular bubble

97. From a map, the distance from a control station to a distant mountain is calculated to be 18.4 mi. The earth curvature correction for this angle line is most nearly

D

(A) 20 ft

(B) 23 ft

(C) 200 ft

(D) 230 ft

98. A data table is shown.

point	BS (ft)	FS (ft)	prelim. elevation (ft)	record elevation (ft)	
BM1	3.57	–	–	100.00	
	2.56	6.61	–	–	96.96
TP1	4.91	5.89	–	–	93.63
	3.33	4.67	–	–	93.87
BM2	–	6.72	–	90.60	90.48

The adjusted (balanced) elevation of TP1 is most nearly

C

(A) 93.57 ft

(B) 93.63 ft

(C) 93.69 ft

(D) 106.37 ft

99. A traverse closes to 1/84,000. The lengths of the traverse sides are given in the table.

traverse side	length (ft)
AB	10,256
BC	8234
CD	4744
DA	12,399

The traverse misclose on the ground is most nearly

C

(A) 0.18 ft

(B) 0.38 ft

(C) 0.42 ft

(D) 0.65 ft

100. In a leveling field book, the correct procedure for checking the computation of elevations is to   D

(A) compute the elevation differences between successive points and compare these with the backsights minus the foresights

(B) compare the average of all the computed elevations with the sum of the backsights minus the sum of the foresights

(C) repeat the computations

(D) sum the backsights, subtract the sum of the foresights, and compare this value with the elevation difference between the end point and the start point

101. For the triangle shown, course LM bears N 13°20′40″ W, and course LN bears N 39°59′10″ E. Side n is 1298 ft and side m is 1487 ft.

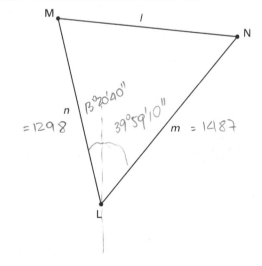

The area of the triangle is most nearly

C

(A)   576,300 ft^2

(B)   578,200 ft^2

(C)   774,100 ft^2

(D)   1,152,700 ft^2

102. A circular curve is segmented as shown.

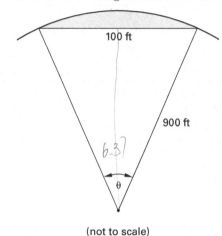

(not to scale)

The area bounded by the circular curve and the 100 ft line is most nearly

B $r = \dfrac{LC}{2\sin\frac{\theta}{2}}$ $A_s = \dfrac{\theta \pi r^2}{360}$

(A)   87.5 ft^2

(B)   92.5 ft^2

(C)   94.5 ft^2

(D)   98.5 ft^2

$A_t = \dfrac{r^2 \sin\theta}{2}$

103. The field data for backsight (BS), foresight (FS), and offset from a cross-sectional survey of a trapezoidal channel are given.   B

point	BS (ft)	FS (ft)	offset (ft)
BM1 (elevation = 502.0 ft)	7.3		
		3.9	0
		8.9	5
		8.9	20
		3.9	25

Most nearly, the area of the channel is

(A)   78.0 ft^2

(B)   100 ft^2

(C)   146 ft^2

(D)   178 ft^2

104. The end cross-sectional areas on a 50.0 ft length of highway are fill 98.24 ft^2 and fill 110.43 ft^2. A prismoidal volume is required, and the computed midway section cross-section area is fill 106.59 ft^2. Using the prismoidal formula, the volume of fill between the end sections is most nearly

C

(A)   190 yd^3

(B)   195 yd^3

(C)   196 yd^3

(D)   216 yd^3

105. Measured highway cross sections S_1 and S_2 are 55.0 ft apart and have areas of 127.22 ft^2 and 187.56 ft^2, respectively. Most nearly, the volume between the sections generated by the end-area method is

(A)   191 yd^3 B

(B)   321 yd^3

(C)   642 yd^3

(D)   8660 yd^3

106. The standard method of computing the volume of a proposed dam basin from a contour map is to   D

(A)   run longitudinal sections across the basin, determine the section areas, and use the end-area method to compute the volume

(B)   mark random points on each contour and use these to form a triangulated irregular network (TIN) file, from which the volume may be determined

(C)   superimpose a grid over the basin, compute the volume below each grid square, and sum these volumes to obtain a total volume of the basin

(D)   determine the area within each contour covering the basin and use the end-area method for the volume computation

107. Three contour lines are depicted, with points A and B positioned as shown.

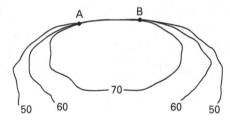

The convergence between points A and B represents

(A) an overhang

(B) a vertical cliff

(C) a ridge

(D) a physical impossibility

108. A cross section is shown.

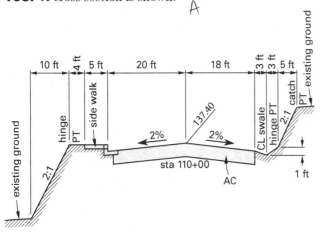

Which set of survey field data corresponds to the cross section?

(A)

toe (ft)	top (ft)	BOW (ft)	TC (ft)	EP (ft)
−39	−29	−25	−20	−18
132.91	137.91	137.83	137.71	137.04

CL (ft)	EP (ft)	FL (ft)	toe (ft)	top (ft)
0	18	21	24	29
137.40	137.04	136.04	137.04	139.54

(B)

top (ft)	toe (ft)	FL (ft)	EP (ft)	CL (ft)
−29	−24	−21	−18	0
139.54	137.04	136.04	137.04	137.40

EP (ft)	TC (ft)	BOW (ft)	top (ft)	toe (ft)
18	20	26	30	40
137.04	137.71	137.83	137.91	132.91

(C)

toe (ft)	top (ft)	BOW (ft)	EP (ft)	EP (ft)
−39	−29	−25	−22	−20
132.91	137.91	137.83	137.71	137.71

CL (ft)	EP (ft)	FL (ft)	top (ft)	toe (ft)
0	18	21	24	29
137.40	137.04	136.04	137.04	139.54

(D)

EG (ft)	CP (ft)	HP (ft)	TC (ft)	CL (ft)
−45	−40	−30	−20	0
132.81	132.91	137.91	137.74	137.40

EP (ft)	CL (ft)	HP (ft)	CP (ft)	EG (ft)
18	21	24	29	34
137.04	136.04	137.04	139.54	139.64

109. On a survey on the North American Datum of 1983 (NAD 83), the grid scale factor is 0.999967, the sea-level scale factor is 0.999351, and measured distance AB is 8456.9 ft. Distance AB on the state-plane grid is most nearly equal to

(A) 8451.1 ft

(B) 8451.7 ft

(C) 8462.1 ft

(D) 8462.7 ft

110. The forward overlap on a pair of adjacent 9 in by 9 in frame aerial photos is 60% and the photo scale is 1 in:400 ft. The distance on the ground between the photo centers is most nearly

(A) 160 ft

(B) 240 ft

(C) 1440 ft

(D) 2160 ft

Practice Exam Answer Keys

Practice Exam 1 Answer Key

#	Ans		#	Ans		#	Ans
1.	A		20.	C		39.	D
2.	A		21.	A		40.	B
3.	A		22.	A		41.	C
4.	B		23.	D		42.	A
5.	D		24.	B		43.	B
6.	A		25.	B		44.	C
7.	A		26.	C		45.	B
8.	D		27.	C		46.	B
9.	B		28.	A		47.	D
10.	D		29.	C		48.	D
11.	B		30.	D		49.	B
12.	B		31.	C		50.	A
13.	D		32.	C		51.	B
14.	D		33.	C		52.	C
15.	A		34.	C		53.	A
16.	C		35.	B		54.	C
17.	C		36.	A		55.	C
18.	B		37.	B			
19.	B		38.	D			

Practice Exam 2 Answer Key

56.	(A) (B) ● (D)	75. (A) (B) (C) ●	94. (A) ● (C) (D)
57.	(A) (B) ● (D)	76. (A) ● (C) (D)	95. (A) ● (C) (D)
58.	(A) (B) (C) ●	77. ● (B) (C) (D)	96. (A) (B) ● (D)
59.	(A) (B) ● (D)	78. ● (B) (C) (D)	97. (A) (B) (C) ●
60.	(A) (B) ● (D)	79. (A) ● (C) (D)	98. (A) (B) ● (D)
61.	(A) (B) (C) ●	80. (A) ● (C) (D)	99. (A) (B) ● (D)
62.	(A) (B) ● (D)	81. (A) ● (C) (D)	100. (A) (B) (C) ●
63.	● (B) (C) (D)	82. (A) (B) ● (D)	101. (A) (B) ● (D)
64.	(A) (B) ● (D)	83. (A) ● (C) (D)	102. (A) ● (C) (D)
65.	● (B) (C) (D)	84. (A) (B) ● (D)	103. (A) ● (C) (D)
66.	(A) (B) ● (D)	85. (A) ● (C) (D)	104. (A) (B) ● (D)
67.	● (B) (C) (D)	86. ● (B) (C) (D)	105. (A) ● (C) (D)
68.	(A) ● (C) (D)	87. (A) (B) (C) ●	106. (A) (B) (C) ●
69.	(A) ● (C) (D)	88. (A) (B) ● (D)	107. (A) ● (C) (D)
70.	(A) (B) (C) ●	89. (A) (B) ● (D)	108. ● (B) (C) (D)
71.	(A) (B) ● (D)	90. ● (B) (C) (D)	109. ● (B) (C) (D)
72.	(A) ● (C) (D)	91. (A) ● (C) (D)	110. (A) (B) ● (D)
73.	(A) ● (C) (D)	92. (A) (B) (C) ●	
74.	(A) ● (C) (D)	93. ● (B) (C) (D)	

Solutions
Practice Exam 1

1. Control points for aerial mapping projects are placed in predetermined positions. The spacing between the control points along each flight line is a function of the contour interval. Most of the control points are placed on the lateral overlap between adjacent flight lines. The control point positions are therefore determined by the number of photos and the contour interval.

The answer is (A).

2. A radial traverse would be more accurate than a closed traverse since each point is measured directly from the known central point instead of being part of a loop. A radial traverse involves one fewer observation than does a closed traverse. Intervisibility is not a function of the type of traverse but depends on the placement of points. Since a radial traverse point is coordinated by a single side shot (bearing and distance), it is essentially unchecked, as opposed to a point on a closed traverse, which is on a closed loop and is therefore checked.

The answer is (A).

3. A benchmark is only associated with elevation and is defined as any point with a known elevation. A mean sea-level tidal point is a benchmark, but it does not fit a general definition.

The answer is (A).

4. A digital elevation model is a series of northing, easting, and elevation points that defines the terrain surface. It is the modern digital alternative to a contour map.

A digital terrain model is a digital map that includes contours and all vector details, such as road centerlines and sidewalks.

A triangulated irregular network (TIN) file is a set of points for which the northing, easting, and elevation are known, and which has been ordered into a series of triplets so that all triplets form a continuous terrain surface.

A triangulated network is another name for a TIN file.

The answer is (B).

5. Errors due to collimation occur when the backsight and foresight distances are not equal. These errors do not occur in barometric or trigonometric leveling. Profile leveling is usually associated with leveling along a route of some length, not at a site such as a river crossing. In reciprocal leveling, collimation errors can be eliminated by computing the mean of the two elevation differences as observed from the two sides of the river.

The answer is (D).

6. The sum of the foresights minus the sum of the backsights is a check normally applied in differential leveling. It is equal to the difference in elevation between the starting point and the ending point. Since these are the same point in a loop, the answer is zero.

The answer is (A).

7. The change in horizontal distance PR is

$$\Delta \text{elev}_{PR} = BS - FS$$
$$= 4.67 \text{ ft} - 7.22 \text{ ft}$$
$$= -2.55 \text{ ft}$$

Use this to find the grade of course PR.

$$g_{PR} = \left(\frac{\Delta \text{elev}_{PR}}{D_{PR}} \right) \times 100\%$$
$$= \left(\frac{-2.55 \text{ ft}}{50 \text{ ft}} \right) \times 100\%$$
$$= -5.1\%$$

The answer is (A).

8. Determine the HI between points A and B.

$$HI_{AB} = \text{elev}_A + BS_A$$
$$= 1000.00 \text{ ft} + 3.34 \text{ ft}$$
$$= 1003.34 \text{ ft}$$

The HI between points B and C is

$$HI_{BC} = HI_{AB} - FS_B + BS_B$$
$$= 1003.34 \text{ ft} - 2.89 \text{ ft} + 5.00 \text{ ft}$$
$$= 1005.45 \text{ ft}$$

The answer is (D).

9. Calculate the grade of line LM.

$$D_{\text{LM}} = (\text{sta M}) - (\text{sta L})$$
$$= (17 + 09.08 \text{ sta}) - (15 + 34.86 \text{ sta})$$
$$= 174.22 \text{ ft}$$
$$\Delta h_{\text{LM}} = \text{FS} - \text{BS}$$
$$= 8.10 \text{ ft} - 3.78 \text{ ft}$$
$$= 4.32 \text{ ft}$$
$$g_{\text{LM}} = \frac{\Delta h_{\text{LM}}}{D_{\text{LM}}} \times 100\%$$
$$= \frac{4.32 \text{ ft}}{174.22 \text{ ft}} \times 100\%$$
$$= 2.48\% \quad (2.5\%)$$

The answer is (B).

10. The height difference between the roof edge and the line of sight is $12.00 \text{ ft} - 4.89 \text{ ft} = 7.11 \text{ ft}$. Therefore, the roof edge is 7.11 ft above the line of sight.

The height difference between the line of sight and the benchmark is equal to the rod reading of 7.90 ft. Therefore, the line of sight is 7.90 ft above the benchmark. The sum of these two height differences, 7.11 ft + 7.90 ft, or 15.01 ft, equals the height difference between the benchmark and the roof edge.

The answer is (D).

11. The height of the instrument is calculated as

$$\text{HI} = \text{elev}_{\text{B}} + \text{RR} - \text{elev}_{\text{A}}$$
$$= 932.1 \text{ ft} + 9.1 \text{ ft} - 937.4 \text{ ft}$$
$$= 3.8 \text{ ft}$$

The answer is (B).

12. Leveling of the highest precision is called geodetic or precise leveling. A geodetic leveling rod consists of an Invar® strip held in a casing of well-seasoned wood or an extruded aluminum alloy. The readings are marked on the Invar® strip.

The answer is (B).

13. NGVD 29 is an older system and is less frequently used. NAD 27 and NAD 83 are not elevation datums but are state-plane coordinate systems. The most common elevation datum for modern surveys is NAVD 88.

The answer is (D).

14. All measured distances must be reduced to sea level. Distances measured at an elevation above sea level are longer than their sea-level equivalents; a negative correction must be applied. All measured distances must be corrected by a grid scale factor before undertaking any computations on a state-plane coordinate system.

The answer is (D).

15. NAVD 88 was instated as the official North American datum in 1993. ASVD02, NMVD03, VIVD09 are geodetic vertical datums in United States territories.

The answer is (A).

16. The horizontal control datum used for the United States is the California Coordinate System (CCS). NAD 29 and NAD 88 do not exist. NAD 27 was superseded by NAD 83 in 1995.

The answer is (C).

17. A photo image that is a true map is known as a digital orthophoto. Usually, every pixel also has an elevation above sea level. A digital orthophoto can be compiled from aerial photos or satellite images.

The answer is (C).

18. The method of contours directly drawn in stereo from the overlap is rarely used today, although it was the standard method in the 1950s and 1960s. Computer interpolation from stereo profiles and automated contouring by image matching are sometimes used today, but the standard method is to generate contours by computer interpolation based on a digital elevation model manually compiled on a stereoplotter.

The answer is (B).

19. A standard aerial photo is 9 in by 9 in. From the photo scale, 1 in on the photo represents 500 ft on the ground. Therefore, 9 in on the photo equals (9 in)(500 ft/in), or 4500 ft, on the ground.

The length of a side is 60% of the 9 in distance. This can be calculated as

$$L_{60\%} = (0.6)(4500 \text{ ft}) = 2700 \text{ ft}$$

The area covered by the overlapping photos is

$$A = \frac{(4500 \text{ ft})(2700 \text{ ft})}{\left(5280 \dfrac{\text{ft}}{\text{mi}}\right)^2} = 0.44 \text{ mi}^2$$

The answer is (B).

20. The parameters to orient a stereo model are three rotations, three translations, and a uniform scale change. Two of the rotations (tip and tilt) are solved using three height points, which must not be in a straight line. The azimuth rotation and scale change are solved using two plan points. Once the model is rotated and scaled, any of the three height points and any of the two plan points can be used to solve for the three translations.

The answer is (C).

21. Find the latitude difference, dy, and departure difference, dx.

$$dx = D \sin\alpha$$
$$dy = D \cos\alpha$$

Compute the coordinates of points B, C, D, and E by adding the latitude and departure differences, dy and dx, which are in italic in the table. For example,

$$dy_{AB} = 60.038$$
$$y_B = 560.000 + 60.038$$
$$dx_{AB} = 389.996$$
$$x_B = 770.000 + 389.996$$

side	distance (ft)	azimuth	point	coordinate y	coordinate x
			A	560.00	770.00
AB	394.59	81°14′54″		*60.038*	*389.996*
			B	620.038	1159.996
BC	292.75	172°08′50″		*−290.005*	*39.998*
			C	330.033	1199.994
CD	332.87	212°44′26″		*−279.986*	*−180.028*
			D	50.047	1019.966
DE	323.88	351°07′42″		*320.005*	*−49.949*
			E	370.052	970.017
				(370.05)	(970.02)

The answer is (A).

22. Since MP is due west, the y coordinate of point P is

$$y_P = y_M = 638.78$$

LN is due north; therefore,

$$D_{LN} = y_M - y_L$$
$$= 638.78 - 540.55$$
$$= 98.23$$

Since bearing LP is N 68°35′00″ W, α is 68°35′00″.

$$D_{PN} = LN \tan\alpha = 98.23 \tan 68°35′00″$$
$$= 250.44$$

The x coordinate of point P is then

$$x_P = x_L - D_{PN}$$
$$= 879.01 - 250.44$$
$$= 628.57$$

The answer is (A).

23. A bearing/bearing intersection is required to compute (y_P, x_P).

$$y_P = \frac{(x_A - x_B) - y_A \tan\alpha_{AP} + y_B \tan\alpha_{BP}}{\tan\alpha_{BP} - \tan\alpha_{AP}}$$

$$= \frac{\begin{array}{c}(2546.09 - 2870.06) - 1675.24\tan 95°34′20″ \\ + 1294.39\tan 48°00′00″\end{array}}{\tan 48°00′00″ - \tan 95°34′20″}$$

$$= 1609.49$$

$$x_P = (y_P - y_A)\tan\alpha_{AP} + x_A$$
$$= (1609.49 - 1675.24)\tan 95°34′20″ + 2546.09$$
$$= 3220.02$$

The answer is (D).

24. The horizontal distance, HD, is given by

$$HD = SD\cos\beta$$

Therefore,

$$HD_{true} = (257.56 \text{ ft})\cos 3°00′00″$$
$$= (257.56 \text{ ft})(0.998630)$$
$$= 257.21 \text{ ft}$$

$$HD_{incorrect} = (257.56 \text{ ft})\cos 4°00′00″$$
$$= (257.56 \text{ ft})(0.997564)$$
$$= 256.93 \text{ ft}$$

The error is determined from the difference in these values.

$$HD_{error} = HD_{true} - HD_{incorrect}$$
$$= 257.21 \text{ ft} - 256.93 \text{ ft}$$
$$= 0.28 \text{ ft}$$

The answer is (B).

25. The area of the entire lot is

$$(6 \text{ ac})\left(\frac{43{,}560 \text{ ft}^2}{1 \text{ ac}}\right) = \left(\frac{784.00 \text{ ft} + AD}{2}\right)(497.00 \text{ ft})$$

$$AD = 267.75 \text{ ft}$$
$$AC = 784.00 \text{ ft} - 267.75 \text{ ft}$$
$$= 516.25 \text{ ft}$$
$$AB = \sqrt{AC^2 + AB^2}$$
$$= \sqrt{497 \text{ ft}^2 + 516.25 \text{ ft}^2}$$
$$= 716.61 \text{ ft} \quad (720 \text{ ft})$$

The answer is (B).

26. Use a right triangle to solve for line GH.

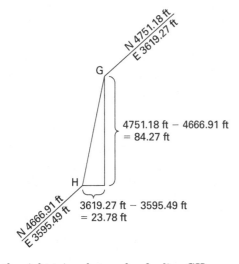

Using the right triangle to solve for line GH,

$$GH = \sqrt{(84.27 \text{ ft})^2 + (23.78 \text{ ft})^2}$$
$$= 87.56 \text{ ft} \quad (87.6 \text{ ft})$$

The answer is (C).

27. From the law of cosines,

$$b^2 = a^2 + c^2 - 2ac \cos B$$

Rearranging to solve for the cosine of B,

$$\cos B = \frac{a^2 + c^2 - b^2}{2ac}$$
$$= \frac{(330.56 \text{ ft})^2 + (380.02 \text{ ft})^2 - (210.90 \text{ ft})^2}{(2)(330.56 \text{ ft})(380.02 \text{ ft})}$$
$$= 0.832699$$

Therefore,

$$\text{angle B} = \arccos 0.832699$$
$$= 33°37'23''$$

The answer is (C).

28. A right triangle is shown.

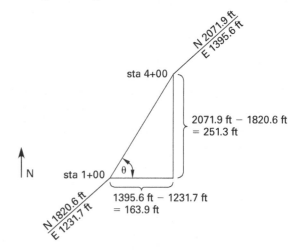

Use the right triangle to solve for the distance and angle between sta 1+00 and sta 4+00.

$$\tan \theta = \frac{\text{opposite side}}{\text{adjacent side}} = \frac{251.3 \text{ ft}}{163.9 \text{ ft}}$$
$$\theta = 56.89°$$
$$= 56°53'24''$$

To get the north to east direction,

$$\text{direction} = 90° - \theta$$
$$= 90° - 56°53'24'$$
$$= 33°06'39' \quad (33°07')$$
$$\text{bearing} = \text{N } 33°06'39' \text{ E} \quad (\text{N } 33°07' \text{ E})$$

The answer is (A).

29. The value of angle EBD is the sum of the values calculated for angle ABE and angle DBC.

$$
\begin{aligned}
\text{ABE} &= 180° - (70°15' + 48°20') \\
&= 61°25' \\
\text{DBC} &= 180° - (55°30' + 65°50') \\
&= 58°40' \\
\text{EBD} &= 180° - (61°25' + 58°40') \\
&= 59°55'
\end{aligned}
$$

The answer is (C).

30. Find the deflection angle, I, at CT.

$$
\begin{aligned}
\tan I = \frac{x}{T} &= \frac{19.464 \text{ ft}}{278.346 \text{ ft}} \\
&= 0.069927 \\
I = \arctan 0.069927 \\
&= 4°
\end{aligned}
$$

The deflection angle is half the central angle, Δ, so the central angle is

$$
\begin{aligned}
\Delta = 2I &= (2)(4°) \\
&= 8°
\end{aligned}
$$

The answer is (D).

31. Use the horizontal curve to solve for x and y.

$$
\begin{aligned}
x = R \sin \alpha \\
&= (500 \text{ ft}) \sin 30°15' \\
&= 251.89 \text{ ft} \\
y = R(1 - \cos \alpha) \\
&= (500 \text{ ft})(1 - \cos 30°15') \\
&= 68.08 \text{ ft}
\end{aligned}
$$

The (x, y) coordinates of point EC are

$$
\begin{aligned}
\text{EC}_x &= 100.4 \text{ ft} + 251.89 \text{ ft} \\
&= 352.29 \text{ ft} \quad (352.3 \text{ ft}) \\
\text{EC}_y &= 92.1 \text{ ft} - 68.08 \text{ ft} \\
&= 24.02 \text{ ft} \quad (24.0 \text{ ft})
\end{aligned}
$$

The answer is (C).

32. The basic equation of a parabolic vertical curve to find the beginning of vertical curve (BVC) is

$$
\text{elev} = \text{elev}_{\text{BVC}} + g_1 x + \left(\frac{r}{2}\right) x^2
$$

$$
\text{elev}_{\text{BVC}} = \text{elev}_{\text{PVI}} - g_1 \left(\frac{L}{2}\right)
$$

$$
\begin{aligned}
&= 310.50 \text{ ft} - (-1.5\%)\left(\frac{7 \text{ sta}}{2}\right) \\
&= 315.75 \text{ ft}
\end{aligned}
$$

The rate of change of grade is

$$
\begin{aligned}
r &= \frac{g_2 - g_1}{L} \\
&= \frac{2.0\% - (-1.5\%)}{7 \text{ sta}} \\
&= 0.5\% / \text{sta}
\end{aligned}
$$

The station of the beginning of vertical curve is

$$
\begin{aligned}
\text{sta}_{\text{BVC}} &= \text{sta}_{\text{PVI}} - \frac{L}{2} \\
&= 8.00 \text{ sta} - \frac{7.00 \text{ sta}}{2} \\
&= 4.5 \text{ sta} \quad (\text{sta } 4+50)
\end{aligned}
$$

Distance BVC to sta 7+10 in units of 100 ft is

$$
\begin{aligned}
D &= \frac{\text{sta}_{7+10} - \text{sta}_{\text{BVC}}}{100} \\
&= \frac{710 \text{ ft} - 450 \text{ ft}}{100 \text{ ft}} \\
&= 2.60
\end{aligned}
$$

The elevation at sta 7+10, y, is

$$
\begin{aligned}
y &= 315.75 \text{ ft} + (-1.5\%)(2.60) \\
&\quad + \left(\frac{0.5 \frac{\%}{\text{sta}}}{2}\right)(2.60)^2 \\
&= 313.54 \text{ ft}
\end{aligned}
$$

The answer is (C).

33. Find where the curve crests.

$$
x = \frac{-g_1}{r}
$$

$$
g_1 = 1.0\%
$$

$$
\begin{aligned}
r &= \frac{g_2 - g_1}{L} = \frac{-1.75\% - 1.0\%}{4 \text{ sta}} \\
&= -0.6875\%/\text{sta}
\end{aligned}
$$

$$x = \frac{-1.0\%}{\frac{-0.6875\%}{\text{sta}}}$$
$$= 1.4545 \text{ sta from BVC}$$

The curve crests at $(\text{sta } 33+00) + (\text{sta } 1+45.45)$, or sta 34+45.45.

The answer is (C).

34. The offset can be found from

$$x = (r_1 + r_2)(1 - \cos I)$$
$$= (1500 \text{ ft} + 1200 \text{ ft})(1 - \cos 15°20'00'')$$
$$= 96.11 \text{ ft}$$

The answer is (C).

35. Convert angles to decimal degrees.

$$15°05' = 15° + \left(\frac{\frac{5'}{60'}}{1°}\right)$$
$$= 15.083°$$

$$11°45' = 11° + \left(\frac{\frac{45'}{60'}}{1°}\right)$$
$$= 11.750°$$

Next, convert angles to radians.

$$15.083° = \frac{15.083°(\pi \text{ rad})}{180°}$$
$$= 0.26325 \text{ rad}$$

$$11.750° = \frac{11.750°(\pi \text{ rad})}{180°}$$
$$= 0.20508 \text{ rad}$$

The length of the compound curve is

$$L = \Delta_1 r_1 + \Delta_2 r_2$$
$$= (0.2633 \text{ rad})(2000 \text{ ft}) + (0.2051 \text{ rad})(2450 \text{ ft})$$
$$= 1028.94 \text{ ft}$$

The answer is (B).

36. Using tangent offsets,

$$x = R \sin \alpha$$
$$100 \text{ ft} = (4000 \text{ ft}) \sin \alpha$$
$$\sin \alpha = \frac{100 \text{ ft}}{4000 \text{ ft}} = \frac{1}{40}$$

Solve for α.

$$\alpha = \arcsin \frac{1}{40} = 1.433°$$
$$y = R(1 - \cos \alpha) = (4000 \text{ ft})(1 - \cos 1.433°)$$
$$= 1.25 \text{ ft} \quad (1.3 \text{ ft})$$

The answer is (A).

37. A construction staking survey lays out and identifies the horizontal and vertical location of proposed improvements prior to the beginning of construction. The contractor uses the field controls placed by the surveyor to place the improvements correctly. A topographical survey is done before design. Design field changes can take place during construction as needed, but the majority of a construction staking survey is performed before construction begins. An as-built survey is done after construction.

The answer is (B).

38. Point K is the reference marker. Each entry under the double strike line indicates the elevation difference from point J to the designed elevation, which is the horizontal distance in feet from the reference marker, point K. The second entry indicates that the targeted location is 25 horizontal ft from point K and 1 ft below the elevation of point K. "EP" is the abbreviation for edge of pavement.

The answer is (D).

39. Point K is the reference marker on the ground. Each entry under the double strike line indicates the elevation difference from point J to the designed elevation, which is the horizontal distance in feet from the reference marker, point K. The first entry indicates the targeted location is 20 horizontal ft from point K and 1.1 ft below the elevation of point K. CP is the abbreviation for catch point of a slope.

The answer is (D).

40. Aerial mapping control is often surveyed using the global positioning system (GPS). Aerial mapping projects cover large areas, and GPS is particularly suited to this application, since intervisibility between points is not required. Triangulation and stadia are obsolete methods, and total station surveying is best suited to smaller projects or to projects in built-up areas where

GPS might not be practical because of interference by tall buildings and trees.

The answer is (B).

41. The corrected length can be found using the equation

$$L_c = L + C_l$$
$$= L + (T_l - T)kL$$

The three corrected lengths are

$$L_{C_1} = 597.410 \text{ ft} + (83°\text{F} - 68°\text{F})$$
$$\times \left(6.45 \times 10^{-6} \frac{1}{°\text{F}}\right)(597.410 \text{ ft})$$
$$= 597.468 \text{ ft}$$
$$L_{C_2} = 597.320 \text{ ft} + (98°\text{F} - 68°\text{F})$$
$$\times \left(6.45 \times 10^{-6} \frac{1}{°\text{F}}\right)(597.320 \text{ ft})$$
$$= 597.436 \text{ ft}$$
$$L_{C_3} = 597.500 \text{ ft} + (49°\text{F} - 68°\text{F})$$
$$\times \left(6.45 \times 10^{-6} \frac{1}{°\text{F}}\right)(597.500 \text{ ft})$$
$$= 597.444 \text{ ft}$$

The answer is (C).

42. The standard equations for distances (in ft) are

$$C_f = 0.0239\left(\frac{D}{1000}\right)^2$$
$$R_f = -0.0033\left(\frac{D}{1000}\right)^2$$

Convert the distance of the 8.6 km line to ft.

$$D_{\text{ft}} = (8.6 \text{ km})\left(3280.8 \frac{\text{ft}}{\text{km}}\right)$$
$$= 28,215 \text{ ft}$$

The corrections for earth curvature and refraction can be determined. The refraction component is negative and opposite in sign to the earth curvature correction.

$$C_f = (0.0239)\left(\frac{28,215 \text{ ft}}{1000}\right)^2$$
$$= 19.03 \text{ ft}$$

$$R_f = (-0.0033)\left(\frac{28,215 \text{ ft}}{1000}\right)^2$$
$$= -2.63 \text{ ft}$$

combined correction $= C_f + R_f$
$$= 19.03 \text{ ft} + (-2.63 \text{ ft})$$
$$= 16.40 \text{ ft}$$

The answer is (A).

43. Calculate the x misclose.

$$x \text{ misclose} = \text{known } x - \text{computed } x$$
$$= 2459.16 \text{ ft} - 2459.38 \text{ ft}$$
$$= -0.22 \text{ ft}$$

Compute the y misclose.

$$y \text{ misclose} = \text{known } y - \text{computed } y$$
$$= 5211.90 \text{ ft} - 5211.71 \text{ ft}$$
$$= +0.19 \text{ ft}$$

The x and y miscloses are combined to form a diagonal linear misclose.

$$\text{linear misclose} = \sqrt{(x \text{ misclose})^2 + (y \text{ misclose})^2}$$
$$= \sqrt{(-0.22 \text{ ft})^2 + (0.19 \text{ ft})^2}$$
$$= 0.2907 \text{ ft}$$

The linear misclose is converted to an accuracy.

$$\text{accuracy} = 1{:}\frac{\text{traverse length}}{\text{linear misclose}}$$
$$= 1{:}\frac{3250 \text{ ft}}{0.2907 \text{ ft}}$$
$$= 1{:}11{,}180 \quad (1{:}11{,}000)$$

The answer is (B).

44. The sum of the four angles is $359°59'20''$. The corrected sum is $360°00'00''$.

$$\text{misclose} = 360° - \sum 4 \text{ angles}$$
$$= 360° - 359°59'20''$$
$$= 40''$$
$$\text{correction per angle} = \frac{\text{misclose}}{\text{no. of angles}}$$
$$= \frac{40''}{4}$$
$$= 10''$$
$$\text{balanced angle D} = 81°55'50'' + \text{correction}$$
$$= 81°55'50'' + 10''$$
$$= 81°56'00''$$

The answer is (C).

45. Calculate the height of the instrument (HI) and missing elevation values.

The height of instrument is

$$HI = BS + elev$$

The elevation is

$$elev = HI - FS$$

Calculate intermediate values as shown.

station	BS (ft) [+b]	HI (ft)	FS (ft) [−]	elevation (ft)
BM1	4.20	4.20 + 431.20 = 435.40		431.20 (known)
TP1	5.34	5.34 + 431.42 = 436.76	3.98	435.40 − 3.98 = 431.42
TP2	8.45	8.45 + 428.46 = 436.91	8.30	436.76 − 8.30 = 428.46
BM2			7.62	429.37 (known) 436.91 − 7.62 = 429.29 (measured)

The vertical error of closure is calculated by subtracting the known elevation from the measured elevation.

$$\begin{aligned} vertical\ error_{closure} &= elev_{measured} - elev_{known} \\ &= 429.29\ ft - 429.37\ ft \\ &= -0.08\ ft \end{aligned}$$

The answer is (B).

46. According to *Surveying* by Moffitt and Bossler, the four general methods of keeping survey field notes are writing a description of the work performed, creating a sketch showing all numerical values, tabulating the survey's numerical values, and combining all three of these methods.

The answer is (B).

47. The following are rearranged to solve for the area.

$$\frac{A}{\pi R^2} = \frac{I}{360°}$$

Area is calculated as

$$A = \left(\frac{I}{360°}\right)\pi R^2 = \left(\frac{47°}{360°}\right)\pi(1100\ ft)^2$$
$$= 496{,}284\ ft^2$$

Convert this area to acres.

$$A = \frac{496{,}284\ ft^2}{43{,}560\ \frac{ft^2}{ac}} = 11.39\ ac \quad (11.4\ ac)$$

The answer is (D).

48. The area of sector 1 is

$$A = \left(\frac{21\ ft + 34\ ft}{2}\right)(20\ ft)\left(\frac{1\ yd}{3\ ft}\right)^2$$
$$= 61.1\ yd^2$$

The area of the remaining three sectors is

$$A' = \left(\frac{34\ ft + 34\ ft}{2}\right)(25\ ft) + \left(\frac{34\ ft + 24\ ft}{2}\right)$$
$$\times (30\ ft) + \left(\frac{24\ ft + 27\ ft}{2}\right)(17\ ft)$$
$$= (2153.5\ ft^2)\left(\frac{1\ yd}{3\ ft}\right)^2$$
$$= 239.3\ yd^2$$

The area of the cross section is

$$\begin{aligned} A_{cross\ section} &= A + A' = 61.1\ yd^2 + 239.3\ yd^2 \\ &= 300.4\ yd^2 \quad (300\ yd^2) \end{aligned}$$

The answer is (D).

49. Use the three-level formula to find the area.

$$\begin{aligned} A &= \Delta h\left(\frac{D_L + D_R}{2}\right) + W\left(\frac{\Delta h_L + \Delta h_R}{4}\right) \\ &= (3.80\ ft)\left(\frac{38.41\ ft + 35.17\ ft}{2}\right) \\ &\quad + (60.00\ ft)\left(\frac{4.84\ ft + 2.99\ ft}{4}\right) \\ &= 257.25\ ft^2 \quad (260\ ft^2) \end{aligned}$$

The answer is (B).

50. Calculate the volume of the excavation.

$$\begin{aligned} V &= LWD \\ &= (80\ ft)(60\ ft)(5\ ft)\left(\frac{1\ yd^3}{27\ ft^3}\right) \\ &= 888.89\ yd^3 \end{aligned}$$

Determine the total cost of excavation.

$$CE_{total} = V(CE_{yd^3})$$
$$= (888.89 \text{ yd}^3)\left(\frac{\$2.00}{yd^3}\right)$$
$$= \$1777.78$$

The dirt is moved 800 ft, or 8 sta. There is no charge for overhaul of the first five stations. Determine the total cost of overhaul.

$$CO_{total} = V(CO_{yd^3})$$
$$= (888.89 \text{ yd}^3)\left(\$0.40 \frac{yd^3}{sta}\right)(8 \text{ sta} - 5 \text{ sta})$$
$$= \$1066.67$$

The total cost is

$$C_{total} = CE_{total} + CO_{total}$$
$$= \$1777.78 + \$1066.67$$
$$= \$2844.45 \quad (\$2850)$$

The answer is (A).

51. Using the prismoidal formula, the middle section dimensions are obtained by simply averaging the two end section dimensions. A common mistake is to compute the two end cross-sectional areas and then average them.

The answer is (B).

52. There are five contours between elevation 505 and elevation 510. Because the contours are at intervals of 1 ft, each contour represents 1 vertical foot change in elevation. Stake 4 is at elevation 509 and stake 3 is at elevation 506.5.

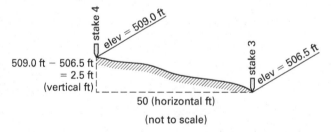

To calculate the grade, divide the vertical change by the horizontal change.

$$grade = \frac{\text{vertical change}}{\text{horizontal change}} = \frac{2.5 \text{ ft}}{50 \text{ ft}} = 0.05$$
$$= 5\%$$

The answer is (C).

53. Option A rises in elevation from sta 1+00, peaks near sta 1+75 to sta 2+00, and decreases to sta 3+00. This follows the values of the EG elevation given in the table.

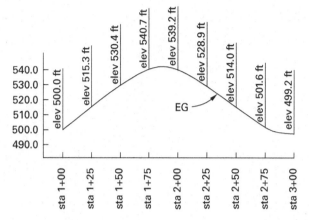

The answer is (A).

54. Calculate the scale.

$$scale = \frac{\text{focal length}}{\text{altitude} - \text{ground elevation}}$$
$$= \frac{6 \text{ in}}{4000 \text{ ft} - 400 \text{ ft}}$$
$$= \frac{1 \text{ in}}{600 \text{ ft}} \quad (1 \text{ in}:600 \text{ ft})$$

The answer is (C).

55. Convert the ground distance and photo distance to the same unit (inches).

$$D_{ground} = (120 \text{ yd})\left(36 \frac{in}{yd}\right) = 4320 \text{ in}$$
$$D_{photo} = (9.144 \text{ mm})\left(\frac{1 \text{ in}}{25.4 \text{ mm}}\right) = 0.36 \text{ in}$$

Calculate the photo scale.

$$\text{photo scale} = 1:\frac{D_{ground}}{D_{photo}} = 1:\frac{4320 \text{ in}}{0.36 \text{ in}}$$
$$= 1:12,000$$

The answer is (C).

Solutions
Practice Exam 2

56. A control survey is the setting out of a network of points that are used as a reference for future surveys. This network can cover an area as large as a state or as small as a few acres.

The answer is (C).

57. The geoid is the surface of mean sea level extended over the entire globe or the surface of the sea if small channels were cut through all the mountains, allowing the sea to flow through all land masses. The geodetic reference surface mentioned in option A is called the spheroid or ellipsoid. The height of mean sea level above or below the spheroid is called the geoidal undulation.

The answer is (C).

58. A control survey is used at the beginning of a project to establish the horizontal reference and vertical reference for the project site. The control survey is used with other surveys, such as construction staking surveys.

The answer is (D).

59. A topographic survey is used to map the land surface and the existing features of a project site. It is typically performed near the beginning of a project prior to engineering design, and it can be done with conventional survey techniques on land or by aerial photogrammetry.

The answer is (C).

60. The elevation of the top of the rebar is

$$\text{elev}_{\text{top of rebar}} = \text{elev}_{\text{BM}} + \text{BS} - \text{FS}$$
$$= 390.66 \text{ ft} + 7.56 \text{ ft} - 1.01 \text{ ft}$$
$$= 397.21 \text{ ft}$$

Since the pipe surface is 1.5 ft below the top of the rebar,

$$\text{elev}_{\text{top of pipe}} = \text{elev}_{\text{top of rebar}} - L_{\text{rebar}}$$
$$= 397.21 \text{ ft} - \left(\frac{18 \text{ in}}{12 \ \frac{\text{in}}{\text{ft}}} \right)$$
$$= 395.71 \text{ ft}$$

The answer is (C).

61. The height of instrument, HI, is

$$\text{HI} = \text{elev}_{\text{BM}} + \text{BS}$$
$$= 2800.20 \text{ ft} + 2.55 \text{ ft}$$
$$= 2802.75 \text{ ft}$$

The elevation of the lath mark plus the foresight, FS, is equal to the HI.

$$\text{HI} = \text{elev}_{\text{lath mark}} + \text{FS}$$

Rearranging to solve for the FS, which is equal to the RR,

$$\text{RR} = \text{FS} = \text{HI} - \text{elev}_{\text{lath mark}}$$
$$= 2802.75 \text{ ft} - 2798.11 \text{ ft}$$
$$= 4.64 \text{ ft}$$

The answer is (D).

62. The elevation of point Q is found from

$$\text{elev}_{\text{Q}} = \text{elev}_{\text{P}} + \left(\sum \text{BS} - \sum \text{FS} \right)$$
$$= 1000.00 \text{ ft}$$
$$\quad + (4.68 \text{ ft} + 3.99 \text{ ft} + 2.07 \text{ ft} + 5.55 \text{ ft})$$
$$\quad - (2.12 \text{ ft} + 1.63 \text{ ft} + 1.64 \text{ ft} + 4.71 \text{ ft})$$
$$= 1006.19 \text{ ft}$$

The answer is (C).

63. The distance from the instrument to point A is

$$x_{\text{A}} = 100(\text{top reading} - \text{bottom reading})$$
$$= (100)(5.15 \text{ ft} - 4.85 \text{ ft})$$
$$= 30.0 \text{ ft}$$

The distance from the instrument to point B is

$$x_{\text{B}} = 100(\text{top reading} - \text{bottom reading})$$
$$= (100)(6.30 \text{ ft} - 5.54 \text{ ft})$$
$$= 76.0 \text{ ft}$$

Calculate the distance AB.

$$x_{AB} = x_B - x_A = 76.0 \text{ ft} - 30.0 \text{ ft}$$
$$= 46.0 \text{ ft}$$
$$\Delta y_{AB} = (\text{height B}) - (\text{height A})$$
$$= (\text{HI} - \text{mid reading to B})$$
$$- (\text{HI} - \text{mid reading to A})$$
$$= (\text{HI} - 5.92 \text{ ft}) - (\text{HI} - 5.00 \text{ ft})$$
$$= -0.92 \text{ ft}$$

The grade of line AB is

$$g_{AB} = \frac{\Delta y_{AB}}{x_{AB}} \times 100\%$$
$$= \frac{-0.92 \text{ ft}}{46.0 \text{ ft}} \times 100\%$$
$$= -2\%$$

The answer is (A).

64. Calculate the difference between the height of point B and point C.

$$y = y_{\text{known}} + \text{BS} - \text{FS}$$
$$y_B = 500.00 \text{ ft} + 4.77 \text{ ft} - 3.65 \text{ ft}$$
$$= 501.12 \text{ ft}$$
$$y_C = 500.00 \text{ ft} + 4.77 \text{ ft} - 5.89 \text{ ft}$$
$$= 498.88 \text{ ft}$$
$$\Delta y_{BC} = 501.12 \text{ ft} - 498.88 \text{ ft}$$
$$= 2.24 \text{ ft}$$

The answer is (C).

65. The elevation of the bottom of the rod (top surface of the pipe) is

$$y_{\text{rod bottom}} = \text{BS} - \text{FS} + y_{\text{BM}}$$
$$= 1.67 \text{ ft} - 8.06 \text{ ft} + 378.54 \text{ ft}$$
$$= 372.15 \text{ ft}$$

Determine the height difference between the center and the outside of the pipe.

$$\Delta y = \frac{5 \text{ ft } 4 \text{ in}}{2} + 1 \text{ in} = 2 \text{ ft } 9 \text{ in}$$

The center of the pipe is 2 ft 9 in (or 2.75 ft) below the top surface of the pipe.

$$y_{\text{pipe center}} = y_{\text{rod bottom}} - \Delta y$$
$$= 372.15 \text{ ft} - 2.75 \text{ ft}$$
$$= 369.40 \text{ ft}$$

The answer is (A).

66. Determine the height of the instrument.

$$\text{HI} = y_A + \text{BS}_A$$
$$= 100.00 \text{ ft} + 4.87 \text{ ft}$$
$$= 104.87 \text{ ft}$$

Determine the elevation of point D by subtracting the height of the intermediate foresight from the height of instrument.

$$y_D = \text{HI} - \text{IFS}_D$$
$$= 104.87 \text{ ft} - 4.77 \text{ ft}$$
$$= 100.10 \text{ ft}$$

The answer is (C).

67. Solve for the height of the instrument (HI).

$$\text{HI} = \text{elev}_{\text{BM1}} - \text{elev}_A + \text{RR}_{\text{BM1}}$$
$$= 1056.6 \text{ ft} - 1059.4 \text{ ft} + 6.2 \text{ ft}$$
$$= 3.4 \text{ ft}$$

Use the HI to solve for the elevation of point B.

$$\text{elev}_B = \text{elev}_A + \text{HI} - \text{RR}_B$$
$$= 1059.4 \text{ ft} + 3.4 \text{ ft} - 3.8 \text{ ft}$$
$$= 1059.0 \text{ ft}$$

The answer is (A).

68. According to California Public Resources Code Chapter 1, Section 8802. California Coordinate System (CCS) has the following definition:

For CCS27, the state is divided into seven zones. For CCS83, the state is divided into six zones. Zone 7 of CCS27, which encompasses Los Angeles County, is eliminated and the area is included in Zone 5 of CCS83. Each zone of CCS27 is a Lambert conformal conic projection based on Clarke's Spheroid of 1866, which is the basis of NAD27. The points of control of zones one to six, inclusive, bear the coordinates: Northing (y) = 000.00 feet and Easting (x) = 2,000,000 feet. The point of control of Zone 7 bears the coordinates: Northing (y) = 4,160,926.74 feet and Easting (x) = 4,186,692.58 feet.

Each zone of CCS83 is a Lambert conformal conic projection based on the Geodetic Reference System of 1980, which is the basis of NAD83. The point of control of each of the six zones bear the coordinates: Northing (y) = 500,000 meters and Easting (x) = 2,000,000 meters. CCS27 or NAD27, NGVD29 and USSD was legislated with US Survey Feet. CCS83 or NAD83 was legislated with meter.

The answer is (B).

69. The North American Datum (NAD) adjustments were horizontal and did not affect the leveling networks. The NAVD 88 adjustment only held one point fixed—

Father Point/Rimouski. NGVD 29 was adjusted to fit to 26 mean sea-level stations.

The answer is (B).

70. The North American Datum of 1927 (NAD 27) is a horizontal datum (not a vertical datum). CAN 91 does not exist. The National Geodetic Vertical Datum of 1929 (NGVD 29) is an older vertical datum. The North American Vertical Datum of 1988 (NAVD 88) is the datum described in the problem statement.

The answer is (D).

71. NAD datums are horizontal datums and are not related to vertical measurements. NAVD 88 is a vertical datum used post-1991. The datum used prior to 1991 was NGVD 29.

The answer is (C).

72. Surveying and mapping problems invariably involve redundant data in order to provide a check on measurements. When this occurs, the correct procedure would be to use the method of least squares to find the unknowns.

The answer is (B).

73. Full control requires sufficient control points to level and scale the stereomodel so that a map can be compiled at an exact scale with contours. Such a stereomodel will have a minimum of three elevation points— one in each corner and two plan points with known position.

The answer is (B).

74. The figure shows the degree of the angle.

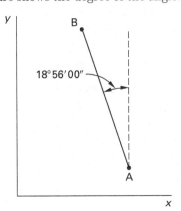

The azimuth of line AB, α, is

$$\alpha = 360°00'00'' - 18°56'00'' = 341°04'00''$$

Find the coordinates of point B.

$$x_B = x_A + D \sin \alpha$$
$$= 310.00 \text{ ft} + (130.65 \text{ ft}) \sin 341°04'00''$$
$$= 267.61 \text{ ft}$$

$$y_B = y_A + D \cos \alpha$$
$$= 275.00 \text{ ft} + (130.65 \text{ ft}) \cos 341°04'00''$$
$$= 398.58 \text{ ft}$$

The answer is (B).

75. This is a bearing/bearing intersection requiring the azimuths BP and AP. If the azimuths are denoted by Az, the coordinates of P may be found from

$$Az_{BP} = 360° - 30°$$
$$= 330°$$
$$Az_{PB} = Az_{BP} - 180°$$
$$= 330° - 180°$$
$$= 150°$$
$$Az_{PA} = Az_{PB} + \alpha$$
$$= 150° + 92°10'00''$$
$$= 242°10'00''$$
$$Az_{AP} = Az_{PA} - 180°$$
$$= 62°10'00''$$

Now find the (x, y) coordinates.

$$y_P = \frac{(x_A - x_B) - y_A \tan Az_{AP} + y_B \tan Az_{BP}}{\tan \alpha_B - \tan \alpha_A}$$
$$= \frac{\begin{array}{c}(500 \text{ ft} - 1000 \text{ ft}) - (1100 \text{ ft}) \tan 62°10'00'' \\ +(1000 \text{ ft}) \tan 330°\end{array}}{\tan 330° - \tan 62°10'00''}$$
$$= 1278.96 \text{ ft}$$

$$x_P = (y_P - y_A) \tan Az_{AP} + x_A$$
$$= (1278.96 \text{ ft} - 1100 \text{ ft}) \tan 62°10'00'' + 500 \text{ ft}$$
$$= 838.94 \text{ ft}$$

The answer is (D).

76. Find distance CP.

$$D_{CP} = \sqrt{(x_P - x_C)^2 + (y_P - y_C)^2}$$
$$= \sqrt{(4400 \text{ ft} - 5000 \text{ ft})^2 + (7700 \text{ ft} - 5000 \text{ ft})^2}$$
$$= 2765.86 \text{ ft}$$

Using the Pythagorean theorem,

$$T = \sqrt{2765.86^2 - 2200^2}$$
$$= 1676.30 \text{ ft}$$

The answer is (B).

77. The difference between the standard temperature of the steel tape and the air temperature is

$$\Delta T = 93°F - 68°F = 25°F$$

The coefficient of thermal expansion of steel, α, is $6.45 \times 10^{-6} /°F$.

$$\text{correction } \Delta L = \alpha L \Delta T$$
$$= \left(6.45 \times 10^{-6} \, \frac{1}{°F}\right)(300.00 \text{ ft})(25°F)$$
$$= 0.05 \text{ ft}$$

At the hotter standard temperature, the tape will expand by 0.05 ft. Therefore, the 300.00 ft mark stands further away than it is from the measurement. For stakeouts, adjustment will be measured length minus $(-)$ the correction. Hence the measured distance between the stakes at 93°F is 300.00 ft $-$ 0.05 ft $=$ 299.95 ft.

The answer is (A).

78. Calculate the elevation of stake B.

$$y_B = y_A + \text{HI} + \text{SD} \sin \alpha - \text{rod reading}$$
$$= 497.26 \text{ ft} + 4.52 \text{ ft} + (187.44 \text{ ft}) \sin 20°45'30''$$
$$\quad -3.05 \text{ ft}$$
$$= 565.16 \text{ ft}$$

The answer is (A).

79. Calculate distance AB.

$$x_{AB} = \sqrt{(x_A - x_B)^2 + (y_A - y_B)^2}$$
$$= \sqrt{\begin{array}{l}(101.56 \text{ ft} - 637.89 \text{ ft})^2 \\ \quad +(556.23 \text{ ft} - 15.33 \text{ ft})^2\end{array}}$$
$$= 761.72 \text{ ft}$$

Therefore, the shortest distance from point A to the arc is

$$\text{shortest distance A to arc} = x_{AB} - R$$
$$= 761.72 \text{ ft} - 700.00 \text{ ft}$$
$$= 61.72 \text{ ft} \quad (61.7 \text{ ft})$$

The answer is (B).

80. Since PQ lies in the first quadrant, its azimuth is the bearing angle, which is 41°57'20''.

$$\text{Az}_{PT} = \text{Az}_{PQ} + \alpha + \beta + \theta$$
$$= 41°57'20'' + 62°10'10'' + 71°44'30''$$
$$\quad +158°32'40''$$
$$= 334°24'40''$$

The bearing angle of PT is

$$\text{bearing angle of PT} = 360° - \text{Az}_{PT}$$
$$= 360° - 334°24'40''$$
$$= 25°35'20''$$

PT lies in the fourth quadrant; therefore, bearing PT is N 25°35'20'' W.

The answer is (B).

81. Calculate the azimuth line CD.

$$\alpha_{AB} = \alpha_{AE} - \text{angle EAB}$$
$$= 111°47'36'' - 51°10'19''$$
$$= 60°37'17''$$
$$\alpha_{BC} = \alpha_{AB} + \text{deflection angle B}$$
$$= 60°37'17'' + 28°00'01''$$
$$= 88°37'18''$$
$$\alpha_{CD} = \alpha_{BC} + 180°00'00'' - \text{angle BCD}$$
$$= 88°37'18'' + 180° - 132°00'00''$$
$$= 136°37'18''$$

The answer is (B).

82. In the figure shown,

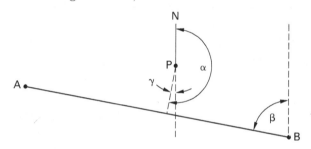

β is 79°34'50'', since bearing BA is N 79°34'50'' W. Therefore,

$$\gamma = 90° - \beta$$
$$= 90° - 79°34'50''$$
$$= 10°25'10''$$

The azimuth of the shortest line from point P to line AB is

$$\alpha = 180°00'00'' + \gamma$$
$$= 180°00'00'' + 10°25'10''$$
$$= 190°25'10''$$

The answer is (C).

83. The first stake is located at sta 144+00, the second stake is located at sta 144+50, and the sixth stake is located at sta 146+50.

The arc distance from the BC to the sixth stake is

$$14,650 \text{ ft} - 14,352 \text{ ft} = 298 \text{ ft}$$

Solve for the interior angle between BC and station 146+50, α, from the BC to the sixth stake.

$$L = R\alpha\left(\frac{\pi}{180}\right)$$

$$\alpha = \frac{L}{R}\left(\frac{180}{\pi}\right)$$

$$= \left(\frac{298 \text{ ft}}{500 \text{ ft}}\right)\left(\frac{180}{\pi}\right)$$

$$= 34.15°$$

Solve for the chord distance, C, from the BC to the sixth stake.

$$C = 2R\left(\sin\frac{\alpha}{2}\right)$$

$$= (2)(500 \text{ ft})\left(\sin\frac{34.15°}{2}\right)$$

$$= 293.62 \text{ ft} \quad (293.6 \text{ ft})$$

The answer is (B).

84. The distance to the nearest point on the curve is illustrated by the distance E.

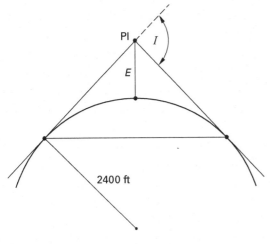

(not to scale)

Distance E can be found from

$$E = r\left(\sec\frac{I}{2} - 1\right)$$

$$= (2400 \text{ ft})\left(\sec\frac{15°30'00''}{2} - 1\right)$$

$$= (2400 \text{ ft})(0.0092)$$

$$= 22.12 \text{ ft}$$

The answer is (C).

85. Determine the total length of curve, L.

$$L_{\text{total}} = \left(\frac{I}{D}\right)L_{\text{between points}}$$

$$= \left(\frac{12°}{2.5°}\right)(100 \text{ ft})$$

$$= 480 \text{ ft} \quad (4.8000 \text{ sta})$$

$$\text{sta}_{\text{EC}} = \text{sta}_{\text{BC}} + L_{\text{total}}$$

$$= 21.0390 \text{ sta} + 4.8000 \text{ sta}$$

$$= 25.8390 \text{ sta}$$

The stationing of BC is sta 21+03.90, and the first point on the curve is sta 22+00. The second point on the curve is therefore sta 23+00.

$$L_{\text{point 2 to EC}} = \text{sta}_{\text{EC}} - \text{sta}_{\text{point 2}}$$

$$= 25.8390 \text{ sta} - 23.0000 \text{ sta}$$

$$= 2.8390 \text{ sta} \quad (283.90 \text{ ft})$$

The central angle generated by a 283.90 ft length of curve is

$$\Delta = DL_{\text{point 2 to EC}} = (2.5°)(2.8390 \text{ sta})$$

$$= 7°05'51''$$

Therefore, the deflection angle at the end of curve is

$$I_{\text{EC}} = \frac{\Delta}{2}$$

$$= \frac{7°05'51''}{2}$$

$$= 3°32'56''$$

The answer is (B).

86. First find the rate of change of grade, r.

$$r = \frac{g_2 - g_1}{L}$$

$$= \frac{2.0\% - (-1.5\%)}{7 \text{ sta}}$$

$$= 0.5 \%/\text{sta}$$

The x value of the lowest point on the curve is

$$x = -\frac{g_1}{r}$$

$$= -\frac{-1.5\%}{0.5 \dfrac{\%}{\text{sta}}}$$

$$= 3 \text{ sta from BVC}$$

The elevation of BVC is

$$\text{elev}_{\text{BVC}} = \text{elev}_{\text{PVI}} - g_1\left(\frac{L}{2}\right)$$

$$= 310.50 \text{ ft} - (-1.5\%)\left(\frac{7 \text{ sta}}{2}\right)\left(100 \frac{\text{ft}}{\text{sta}}\right)$$

$$= 315.75 \text{ ft}$$

The elevation of the lowest point, P, is

$$\text{elev}_{\text{P}} = \text{elev}_{\text{BVC}} + g_1 x + \left(\frac{r}{2}\right)x^2$$

$$= 315.75 \text{ ft} + (-1.5\%)(3 \text{ sta})$$

$$+ \left(\frac{0.5 \frac{\%}{\text{sta}}}{2}\right)(3 \text{ sta})^2$$

$$= 313.50 \text{ ft}$$

The answer is (A).

87. Using the basic equation of a parabolic vertical curve,

$$y = y_{\text{BVC}} + g_1 x + \left(\frac{r}{2}\right)x^2$$

$$y_{\text{BVC}} = y_{\text{PVI}} - g_1\left(\frac{L}{2}\right)$$

$$= 506.98 \text{ ft} - (-2.5\%)\left(\frac{8 \text{ sta}}{2}\right)$$

$$= 516.98 \text{ ft}$$

$$r = \frac{g_2 - g_1}{L} = \frac{1.5\% - (-2.5\%)}{8 \text{ sta}}$$

$$= 0.5\%/\text{sta}$$

Substitute this value into the curve equation.

$$y = 516.98 \text{ ft} + (-2.5x) + \left(\frac{0.5 \frac{\%}{\text{sta}}}{2}\right)x^2$$

$$= 516.98 \text{ ft} - 2.5x + 0.25x^2$$

The answer is (D).

88. First find the offset, x.

$$x = 0.5x_{\text{total}}$$

$$= (0.5)(24.00 \text{ ft})$$

$$= 12.00 \text{ ft}$$

It is known that

$$x = r(1 - \cos I)$$

Therefore,

$$12.00 \text{ ft} = (1000.00 \text{ ft})(1 - \cos I)$$

Solving for the cosine of I,

$$\cos I = 1 - \frac{12.00 \text{ ft}}{1000.00 \text{ ft}}$$

$$= 0.988000$$

Therefore,

$$I = \arccos 0.988000 = 8°53'06''$$

It is known that

$$y = \frac{D}{2} = r \sin I$$

$$= (1000.00 \text{ ft}) \sin 8°53'06''$$

$$= 154.45 \text{ ft}$$

Solving for D yields

$$D = (2)(154.45 \text{ ft})$$

$$= 308.90 \text{ ft}$$

The answer is (C).

89. Calculate the central angle of A and the central angle of B.

$$\text{central angle} = \frac{L}{R}$$

$$\text{central angle of A} = \frac{310 \text{ ft}}{1000 \text{ ft}}$$

$$= 0.310 \text{ rad}$$

$$\text{central angle of B} = \frac{750 \text{ ft}}{2000 \text{ ft}}$$

$$= 0.375 \text{ rad}$$

The total intersection angle is

$$I = \text{central angle of A} + \text{central angle of B}$$

$$= (0.310 \text{ rad} + 0.375 \text{ rad})\left(\frac{180°}{\pi \text{ rad}}\right)$$

$$= 39.2476° \quad (39°14'51'')$$

The answer is (C).

90. The elevation of point BMb is

$$\text{elev}_{\text{BMb}} = \text{elev}_{\text{BMa}} + \text{BS}_1 - \text{FS}_1 + \text{BS}_2$$
$$- \text{FS}_2 + \cdots + \text{BS}_n - \text{FS}_n$$
$$= 783.67 \text{ ft} + 4.67 \text{ ft} - 5.59 \text{ ft}$$
$$+ 3.0 \text{ ft} - 6.51 \text{ ft} + 4.26 \text{ ft}$$
$$- 5.74 \text{ ft} + 3.22 \text{ ft} - 7.38 \text{ ft}$$
$$= 773.60 \text{ ft}$$

The change in elevation from BMb to BMc is

$$\Delta\text{elev}_{\text{BMb}-\text{BMc}} = \text{elev}_{\text{BMc}} - \text{elev}_{\text{BMb}}$$
$$= 763.60 \text{ ft} - 773.60 \text{ ft}$$
$$= -10.00 \text{ ft}$$

The grade of line BMb – BMc can now be found.

$$g_{\text{BMb}-\text{BMc}} = \frac{\Delta\text{elev}_{\text{BMb}-\text{BMc}}}{\text{HD}_{\text{BMb}-\text{BMc}}}$$
$$= \left(\frac{-10.00 \text{ ft}}{200.00 \text{ ft}}\right) \times 100\%$$
$$= -5\%$$

The answer is (A).

91. The rate of change of grade is

$$r = \frac{g_2 - g_1}{L} = \frac{-1\% - 3\%}{6.00 \text{ sta}} = -0.66\%/\text{sta}$$

Calculate the tangent offset.

$$y' = \left(\frac{r}{2}\right)x^2$$
$$= \left(\frac{-0.66 \frac{\%}{\text{sta}}}{2}\right)(2 \text{ sta})^2$$
$$= -1.32 \text{ ft}$$

The answer is (B).

92. When a stake is set on a construction project, it is often accompanied by another stake called a lath (2 ft or 4 ft in length), which contains information about the stake, such as distance to the stake or elevation of the top of the stake.

The answer is (D).

93. The code indicates that the elevation of the top of the stake at point J is 101.00 ft.

The answer is (A).

94. The reference point slope stake (RPSS), is written on a witness stake to indicate that the data on the stake refer to a slope stake some distance away (e.g., 10 ft). The stake next to the witness stake is a reference point.

The answer is (B).

95. The height of instrument measured with a tape only gives the height of the instrument above the ground at the instrument. If a new instrument is set up between each point measured on the section, the height of instrument is not required, but this is never the case in the field. The height of instrument above a datum can be computed by adding the rod reading to the known elevation of a backsight.

The answer is (B).

96. The two-peg test will calibrate the deviation of the line of sight of the level from a level line. The elevation difference between P and Q can be determined twice: first from the backsight/foresight reading when set up at the midway point, and then from the two rod readings taken at the second setup. If these two elevation differences are not equal, the line of sight of the level will be in error.

The answer is (C).

97. The distance from the control station to the mountain is

$$x = (18.4 \text{ mi})\left(5280 \frac{\text{ft}}{\text{mi}}\right) = 97,152 \text{ ft}$$

Determine the earth curvature correction.

$$C_f = (0.0239)\left(\frac{x}{1000}\right)^2$$
$$= (0.0239)\left(\frac{97,152 \text{ ft}}{1000}\right)^2$$
$$= 226 \text{ ft} \quad (230 \text{ ft})$$

The answer is (D).

98. The unadjusted elevation is determined from

$$y_{\text{unadj}} = y_{\text{BM1}} + \text{BS1} - \text{FS1} + \text{BS2} - \text{FS2} + \cdots$$

The unadjusted elevations for test point 1 and benchmark 2 are

$$y_{\text{TP1,unadj}} = 100.00 \text{ ft} + 3.57 \text{ ft} - 6.61 \text{ ft}$$
$$+ 2.56 \text{ ft} - 5.89 \text{ ft}$$
$$= 93.63 \text{ ft}$$
$$y_{\text{BM2,unadj}} = 100.00 \text{ ft} + 3.57 \text{ ft} - 6.61 \text{ ft}$$
$$+ 2.56 \text{ ft} - 5.89 \text{ ft} + 4.91 \text{ ft}$$
$$- 4.67 \text{ ft} + 3.33 \text{ ft} - 6.72 \text{ ft}$$
$$= 90.48 \text{ ft}$$

The misclose is

$$\text{misclose} = y_{BM2,\text{record}} - y_{BM2,\text{unadj}}$$
$$= 90.60 \text{ ft} - 90.48 \text{ ft}$$
$$= 0.12 \text{ ft}$$

Because TP1 is the midpoint of the leveling,

$$y_{TP1,\text{adj}} = y_{TP1,\text{unadj}} + \left(\frac{1}{2}\right)\text{misclose}$$
$$= 93.63 \text{ ft} + \left(\frac{1}{2}\right)(0.12 \text{ ft})$$
$$= 93.69 \text{ ft}$$

The answer is (C).

99. Calculate the traverse misclose on the ground.

$$\frac{1}{84,000} = \frac{\text{misclose}}{\sum \text{side lengths}}$$

Calculate the misclose.

$$\frac{\sum \text{side lengths}}{84,000} = \text{misclose}$$
$$\sum \text{side lengths} = 10,256 \text{ ft} + 8234 \text{ ft}$$
$$+4744 \text{ ft} + 12,399 \text{ ft}$$
$$= 35,633 \text{ ft}$$
$$\text{misclose} = \frac{35,633 \text{ ft}}{84,000}$$
$$= 0.42 \text{ ft}$$

The answer is (C).

100. The universal method of checking the computation of elevations from differential leveling is to subtract the sum of the backsights from the sum of the foresights, and then compare this value with the elevation difference between the first and last point in the level route. Option (C) is a check, but it is not as effective as the method just described. Option (B) is not a check, since it compares an absolute value with differences. Option (A) is similar to option (C), but it is not as effective as option (D).

The answer is (D).

101. Since one bearing is west of north and the other is east of north,

$$\text{angle L} = \text{bearing angle of LM}$$
$$+\text{bearing angle of LN}$$
$$= 13°20'40'' + 39°59'10''$$
$$= 53°19'50''$$

The area of the triangle, A, is

$$A = \left(\frac{mn}{2}\right)\sin \text{L}$$
$$= \left(\frac{(1298 \text{ ft})(1487 \text{ ft})}{2}\right)\sin 53°19'50''$$
$$= 774,071 \text{ ft}^2 \quad \left(774,100 \text{ ft}^2\right)$$

The answer is (C).

102. The area of the triangle bounded by the radii and the 100 ft line is

$$A = \frac{1}{2}bh$$

Using the Pythagorean theorem, the height of the triangle is

$$h = \sqrt{r^2 - \left(\frac{L}{2}\right)^2} = \sqrt{(900 \text{ ft})^2 - \left(\frac{100 \text{ ft}}{2}\right)^2}$$
$$= 898.61 \text{ ft}$$

The area of the triangle is

$$A_t = \frac{1}{2}bh$$
$$= \left(\frac{1}{2}\right)(100 \text{ ft})(898.61 \text{ ft})$$
$$= 44,930.5 \text{ ft}^2$$

The area of the segment bounded by the two radii, A_s, is

$$A_s = \pi r^2\left(\frac{\theta}{360}\right)$$

Determine the angle, θ.

$$\theta = 2\arcsin\left|\frac{\frac{L}{2}}{r}\right|$$
$$= 2\arcsin\frac{\frac{100 \text{ ft}}{2}}{900 \text{ ft}}$$
$$= 6.3695°$$

Therefore,

$$A_s = \pi(900 \text{ ft})^2\left(\frac{6.3695°}{360°}\right)$$
$$= 45,023.2 \text{ ft}^2$$

The area of the segment bounded by the curve and the 100 ft line is

$$A_s - A_t = 45{,}023.2 \text{ ft}^2 - 44{,}930.5 \text{ ft}^2$$
$$= 92.68 \text{ ft}^2 \quad (92.5 \text{ ft}^2)$$

The answer is (B).

103. Calculate the height of instrument (HI) value and elevation values for each offset.

The height of instrument is

$$\text{HI} = \text{BM1} + \text{BS}$$

The elevation is

$$\text{elev} = \text{HI} - \text{FS}$$

Calculate intermediate values as shown.

point	BS (ft)	FS (ft)	offset (ft)	HI (ft)	elevation (ft)
BM1	7.3			502.0 + 7.3 = 509.3	502.0
		3.9	0		509.3 − 3.9 = 505.4
		8.9	5		509.3 − 8.9 = 500.4
		8.9	20		509.3 − 8.9 = 500.4
		3.9	25		509.3 − 3.9 = 505.4

Plot the surveyed points on a graphical representation of the surveyed channel.

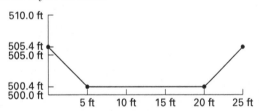

Calculate the area of the channel.

$$A = \left(\frac{b_1 + b_2}{2} \right) h$$
$$= \left(\frac{15 \text{ ft} + 25 \text{ ft}}{2} \right)(5 \text{ ft})$$
$$= 100 \text{ ft}^2$$

The answer is (B).

104. Calculate the volume of the fill.

$$V = \frac{L(A_1 + 4A_m + A_2)}{6}$$
$$= \frac{(50.0 \text{ ft})\left(98.24 \text{ ft}^2 + (4)(106.59 \text{ ft}^2) + 110.43 \text{ ft}^2\right)}{(6)\left(27 \dfrac{\text{ft}^3}{\text{yd}^3}\right)}$$
$$= 196 \text{ yd}^3$$

The answer is (C).

105. The end-area volume is

$$V = \left(\frac{A_{S_1} + A_{S_2}}{2} \right) L$$
$$= \left(\frac{127.22 \text{ ft}^2 + 187.56 \text{ ft}^2}{2} \right)\left(\frac{55.0 \text{ ft}}{27 \dfrac{\text{ft}^3}{\text{yd}^3}} \right)$$
$$= 320.61 \text{ yd}^3 \quad (321 \text{ yd}^3)$$

The answer is (B).

106. All of the options could be used to compute the basin volume. However, option D is preferred, since areas of contour shapes can be quickly determined by x-y digitizing or from planimeter measurements. The end-area formula then yields the volume.

The answer is (D).

107. On a vertical cliff, all contours have the same planimetric position on a map. The section AB, therefore, represents a vertical face or a vertical cliff.

The answer is (B).

108. The point where the proposed improvements meet the existing catch point or toe is located at 39 ft left (−39 ft) of the centerline (CL). The top of slope is located 29 ft left (−29 ft) of CL. The back of walk (BOW) is located 25 ft left (−25 ft) of CL. The top of curb (TC) is located 20 ft left (−20 ft) of CL. The edge of pavement (EP) is located 18 ft left (−18 ft) of CL. The CL is the origin for the offsets, so the offset is shown as zero. The right EP is located 18 ft right of CL. The flowline (FL) is located 21 ft right of CL. The toe of slope is located 24 ft right of CL. The top of slope is located 29 ft right of CL.

The answer is (A).

109. Find distance AB on the state-plane grid from

$$D_{\text{grid}} = D_{\text{measured}}(\text{sea-level scale factor})$$
$$\times (\text{grid scale factor})$$
$$= (8456.9 \text{ ft})(0.999351)(0.999967)$$
$$= 8451.1 \text{ ft}$$

The answer is (A).

110. The overlap is 60% and the distance between the photo centers is $100\% - 60\% = 40\%$. Since the photo is 9 in wide, this represents $(9 \text{ in})(40\%) = 3.6 \text{ in}$.

The photo scale 1 in:400 ft means that 1 in on the photo represents 400 ft on the ground. Therefore, 3.6 in on the photo represents $(3.6 \text{ in})(400 \text{ ft})$, or 1440 ft.

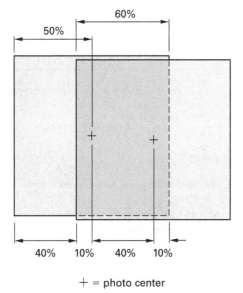

$+$ = photo center

The answer is (C).